AFTER PRESERVATION

AFTER PRESERVATION

SAVING AMERICAN NATURE
IN THE AGE OF HUMANS

EDITED BY

Ben A. Minteer &
Stephen J. Pyne

The University of Chicago Press Chicago and London

BEN A. MINTEER holds the Arizona Zoological Society Chair
in the School of Life Sciences at Arizona State University.
He has published a number of books, including *Refounding
Environmental Ethics* and *The Landscape of Reform*.
STEPHEN J. PYNE is a Regents' Professor in the School of Life
Sciences at Arizona State University. He is the author, editor,
or coeditor of many books, most recently *The Last Lost World*
and *Fire: Nature and Culture*.

The University of Chicago Press, Chicago 60637
The University of Chicago Press, Ltd., London
© 2015 by The University of Chicago
All rights reserved. Published 2015.
Printed in the United States of America

24 23 22 21 20 19 18 17 16 15   1 2 3 4 5

ISBN-13: 978-0-226-25982-6 (cloth)
ISBN-13: 978-0-226-25996-3 (paper)
ISBN-13: 978-0-226-26002-0 (e-book)
DOI: 10.7208/chicago/9780226260020.001.0001

Library of Congress Cataloging-in-Publication Data
After preservation : saving American nature in the age of humans /
edited by Ben A. Minteer and Stephen J. Pyne.
pages cm
Includes bibliographical references and index.
ISBN 978-0-226-25982-6 (cloth : alk. paper) —
ISBN 978-0-226-25996-3 (pbk. : alk. paper) —
ISBN 978-0-226-26002-0 (e-book)
1. Environmental management—United States. 2. Environmental policy—
United States. I. Minteer, Ben A., 1969– II. Pyne, Stephen J., 1949–
GE310 .A38015
363.7'0720973—dc23
2014035525

♾ This paper meets the requirements of ANSI/NISO
Z39.48-1992 (Permanence of Paper)

Contents

Writing on Stone,
Writing in the Wind

The preservationist mission — to shield nature from human manipulation, intrusion, and above all, destruction — has inspired generations of American environmentalists to take action on behalf of threatened species and wild landscapes. The movement's visionaries are revered figures in the tradition, from Henry Thoreau and John Muir, to Aldo Leopold, David Brower, Rachel Carson, and Ed Abbey. The preservationist policy record, too, is widely celebrated and imitated around the world (though often with mixed results): The Wilderness Act (1964) and the Endangered Species Act (1973) have for decades served as legal bulwarks against the march of human development and ecological exploitation.

"Saving American nature," however, has never been easy. The historical record is also strewn with the wreckage of failed campaigns and cratered with ecological losses: the extinction of the passenger pigeon; the damming of Yosemite's Hetch Hetchy Valley; the relentless pressure on American wildlands exerted by countless chainsaws, bulldozers, roads, cities, and summer homes. And even the signature victories — protecting the spotted owl in the Pacific Northwest; keeping the Arctic National Wildlife Refuge closed to oil exploration; saving the California condor from the passenger pigeon's fate — remain tenuous and incomplete. Brower, who did as much as anyone in the tradition (and more than most) to make preservationism a potent political and cultural force in American life in the postwar era liked to say, in fact, that "when they [exploitationists] win, it's forever; when we win, it's merely a stay of execution." The preservationist challenge as he saw it was to maintain a sharp-eyed perseverance in the face of unrelenting environmental threats. If they wanted to ensure a human future rich in wild species and wilderness experiences, Brower implored, preservationists had better be "eternally vigilant."

Yet what exactly are preservationists to be "vigilant" in protecting? We've learned that this is far from a simple question in the time since Brower's zenith in American environmentalism. Rapidly changing biophysical conditions and shifting ecological baselines have at the very least seriously complicated the preservationist agenda. For some, especially for those who view "wild" as the essence of wilderness, it hardly matters because what results may be prompted by human actions but lies beyond human control.

For others these forces have made the very idea of preservation anathema to good conservation stewardship and environmental management in this century. Accelerating climate change, rapid urbanization and suburban sprawl, agricultural intensification, metastasizing fire regimes, the spread of invasive species, and other quickening anthropogenic forces—all appear to be collectively undermining venerable preservationist standards and sentiments. The traditional mantra—"preserve the wilderness!"—no longer seems very useful or compelling to those scientists, conservationists, and environmental writers today who argue that this celebrated object of preservationist passion and political concern is being radically challenged and transformed by global environmental change, most of it driven by humans. If it survives at all, they write, the future of the preservationist model in this "new normal" of global impact and disruption of earth systems will not resemble its past.

Many of these ideas about human influence and environmental control have converged recently around the idea of the "Anthropocene," a term that has migrated to environmentalist

Figure 1. President Theodore Roosevelt and wilderness advocate John Muir at Glacier Point, Yosemite, in 1903. The preservationist tradition has deep roots in American environmental thought and practice, and counts among its ranks an impressive group of writers, scientists, and activists who have long championed the protection of wild species and places on the continent. The philosophical rift between nature protection and a more human-centered agenda, which first became obvious in Muir's time, can still be seen in today's debates over our environmental future in the Anthropocene. Photo credit: Library of Congress Prints and Photographs Division.

circles from its origins in discussions about the naming of the current geological period of human dominance on the planet. Originated by the Nobel Prize–winning atmospheric chemist Paul Crutzen and first promoted to the wider scientific community in an article published in the journal *Nature* in 2002, the term was an attempt to recognize the full suite of planetary changes driven by humans: extensive land transformation, control of the nitrogen cycle, water diversion, and especially anthropogenic alteration of the atmosphere through the emission of greenhouse gases (GHGs). People no longer just talked about the weather: they were changing the climate, although without much control. What had been an environmental fixture against which people acted was now another expression of the human presence. Crutzen argued that these changes, which he traced to the beginnings of the industrial era in the late eighteenth century, mark the beginning of a new geological period, the "Age of Man."

The "Anthropocene" has proved to be a polarizing designation. For some, the notion that humans have become a geological force on the planet—a species able to write its presence in the rocks—is a liberating revelation. It means that we should get on with the business of smart planetary management and get over outmoded myths of a separate, pristine, wild nature that exists free from human influence (and an environmental politics that limits human manipulations of nature). For others, the Anthropocene idea signals the tragic consummation of the destructive human domination of the earth, a last threshold crossed on the march to total ecological despotism. Even for those skeptical of its scientific basis, the Anthropocene remains a reminder of the singular power of humans on the landscape, and stands as a warning of what might lay ahead for us (and for nature) if we do not try to reverse the current course. The Anthropocene, in short, has become an environmentalist Rorschach.

It has in the process exposed deep fault lines and areas of strategic disagreement over the motives, practices, and goals of nature protection in the twenty-first century. Or, to shift meta-phors, nature preservationism today appears to be caught in a cross fire. On one side it continues to be assailed by its familiar political foes; that is, those who argue that the preservation of nature requires unacceptable economic sacrifices and reflects a radical philosophy far outside of the American mainstream. On another it's being raked by the apparently inexorable forces of rapid global change, which are upending preservationist ideals of a nature independent of human influence and impact—and consequently challenging the traditional preservationist ethic minimizing human manipulations of species and landscapes that require more intensive management if they are to survive in an uncertain future.

And on yet another side (perhaps its most exposed and vul-nerable flank), nature preservationism is drawing the fire of a new wave of environmentalists promoting a more anthropo-centric vision of humans and the environment, a pro-growth, often explicitly anti-preservationist politics appropriate to life in the Anthropocene. Arguing that we need a different way of thinking about the human-nature relationship, these "post-preservationist" environmentalists promote a vision in which human interests and needs take center stage and in which we actively embrace our responsibility as shapers and builders of the planetary future. The traditional focus on the wilderness; the knee-jerk hostility to corporate America and distaste for the market; the neglect of working lands and the city—such out-dated preservationist beliefs are roundly rejected by the new Anthropocene-ic environmentalists. They praise instead the phi-losophy of human advancement and the geography and value of nature-culture hybrids—systems in which elements of wildness still persist, yet become inextricably intermingled with human interests, intent, and artifice.

Aldo Leopold once defined a conservationist as someone "who is humbly aware that with each stroke [of an axe] he is writing his signature on the face of his land. Signatures of course differ, whether written with axe or pen, and this is as it should be." And, we might add today, signatures written on stone and in the wind.

* * *

After Preservation brings together a diverse and distinguished set of writers to consider whether and how the American preservationist ideal might survive in an era of expanding human impact on the land, its biota, and the climate they share. The contributors include some of the most prominent environmental scientists, historians, philosophers, environmental writers, journalists, advocates, and policy activists working today. All are known for their serious and sustained engagement with the challenge of American nature preservation, whether as an idea, a historical or philosophical tradition, or as an environmental practice.

We asked for their thoughts in their own voices. In the pages that follow you'll find spirited arguments for a greater human role in environmental systems in the Anthropocene alongside deeply skeptical assessments of the interventionist ethic and the more human-centered vision for nature conservation and environmentalism. But you'll also read attempts to find a middle ground, to reconcile the twin impulses of pragmatism and purity in nature preservation. There are ruminations on the meaning and value of wilderness preservation half a century after the passage of the Wilderness Act, as well as reflections on the management and enduring challenges of conservation, preservation, and restoration on public and private lands. You'll hear meditations on species lost and species saved, and thoughtful reflections on the challenge of coping with a rapidly changing landscape and society. And you'll watch many authors struggling to find solid historical and philosophical footing on the scree field that has become the terrain of preservationism.

We've tried hard not to enforce any specific editorial agenda on the discussion. Our goal instead was to encourage a variety of styles, arguments, and illustrations relevant to the future of nature preservation in the Anthropocene rather than to task writers with a specific writing assignment. We will cop to a modest editorial strategy, however, one that has caused at least a few of our authors to approach the discussion with some trepidation. The title of the book (*After Preservation*) is, we should emphasize, meant to be a provocation. It's a nudge to our authors and readers to think critically and carefully about the viability of the nature preservationist tradition in the "Age of Humans." In other words, we aren't suggesting that the era of preservation is over; we certainly aren't trying to hasten its demise. Maybe Holmes Rolston has it about right, then, with his interrogative rephrasing of our title in his own essay for the book.

In terms of style and structure, you should think of this book as a "symposium" in the classical sense. Or, even better, a "salon." We've invited an influential and thoughtful group of writers to share their ideas about the fate of nature preservation in a humanizing world in a series of short, accessible pieces, often written in a personal style, and prepared for a general audience. Given that our writers were granted free rein to ruminate on the "After Preservation" theme, some chose to focus primarily on the Anthropocene construct, while others adopted an alternative tack, for example, exploring and reconsidering, in more general terms, conservation, preservation, and restoration in this century. We've imposed a domestic frame to the discussion to ground the conversation and anchor it in the American tradition. Still, we think that there are ideas and arguments in these essays that can be scaled up to a larger context, or at least start a more diverse dialogue with other traditions and cultural contexts that might reveal important sympathies and disparities among the American preservationist tradition and other narratives and social practices around the globe.

Finally, we've resisted pigeonholing authors and essays into narrow thematic sections or editorial categories. Still, we think there is a clear logic and rhythm to the cascade of essays that follow; a cadence that we hope will be discernible as you read through them. Among other things, this organization allows you to make your own discoveries, and to consider the interplay of voices and arguments on your terms rather than on ours.

We see ourselves, then, as something akin to stage managers in this production. Our goal has been to keep our own agendas from shaping the course or outcome of the discussions and debates here (beyond our own individual contributions as essayists later in the book). We've created the venue for the conversation and have tried to explain our rationale in putting together this august gathering of environmentalist voices. Our job now is to mostly keep out of the way.

Our final thought appears in the journalist John McPhee's masterful *Coming into the Country*, his account of the people and environment of the upper Yukon region the Alaskan interior. At the end of the book, McPhee, a man with clear and well-known preservationist sympathies (having written on the "Archdruid" David Brower), reflects on his attitude toward the Gelvins, a family of gold prospectors. As he watches them cut into a pristine Alaskan landscape, he neither condemns them nor apologizes for their actions. Instead, he finds himself in a messier and far less absolute moral gamut, one more appropriate to a world in which people and wilderness do and often must collide, in both admirable and destructive ways. "Only an easygoing extremist," McPhee concludes, "would preserve every bit of country. And extremists alone would exploit it all. Everyone else has to think the matter through—choose a point of tolerance, however much the point might tend to one side."

Welcome to the salon.

Ben A. Minteer and Stephen J. Pyne
Tempe, AZ
September 2014

Restoring the Nature of America

Andrew C. Revkin

The deeply disrupted state of American nature, and conservation biology, was never more apparent to me than in 2007, when I was slogging through waterlogged saw grass in Everglades National Park with some government biologists, trying to close in on a radio-tagged 10-foot-long female Burmese python. The scientists' goal was to gain a better understanding of this South Asian snake's haunts and habits in hopes of blunting its spread now that pythons were breeding in balmy Florida thanks to reptile enthusiasts who had discarded overgrown pets.

Questions abounded. Would they make it beyond Florida, particularly in a warming world? Should the giant constrictor still even be called Burmese? How much of the overloaded workday of these park biologists should be spent figuring out how to extirpate this invader? Regulations—along with the freezer back at their lab filled with a dizzying array of park wildlife extracted from pythons' stomachs—made this battle a priority. But there was so much other work to be done: studying less flamboyant invasive species, assessing the park's endangered ones, gauging policies that might restore seasonal freshwater flows, projecting the ecological impact on the park from rising sea levels in a human-heated climate.

The beeps from the handheld antennas of the trackers indicated the snake must be very close. Close indeed. Glancing down, I jumped as I saw its marbled back sliding through the brush inches ahead of my wet shoe. Since then, snake hunts and traps have been tried, but herpetologists largely conclude that this immigrant is here for the long haul.

As with my python reporting, much of what I've seen in 30-plus years of covering the human impact on this planet's veneer of life reveals several things:

· The traditional toolkit of twentieth-century environmental protection is utterly inadequate in considering the biological and social complexities shaping today's and tomorrow's environmental challenges.
· The concept of ecological restoration has lost much of its meaning in the face of the biological Waring blender of global human mobility and trade.
· In many instances, a focus on traits, whether in ecosystems or society, is more apt to pay off than a focus on quantitative goals or hard boundaries.

With all of this in mind, when I consider strategies for "saving American nature in the Age of Humans," I find myself needing to modify some of the terms. "Restoring the nature of America in the Age of Humans" feels better.

I'll explain what I mean shortly. But first I'll explore why I think this semantic adjustment is needed in the first place. For one thing, nature, as some of the other contributors to this volume have articulated for a long time, has never been some separate entity that humans are in a position to save — as in reaching from a boat to snag a struggling swimmer. Ever since *Homo sapiens* evolved and then spread to become a near-planetwide presence and now a powerful climatic and evolutionary influence, nature has included us. Implicit in "saving nature" is saving ourselves. We are all in the same boat.

Then there's the word "saving." It can mean either rescuing or setting aside for future use. Individual species can be saved. But nature generally doesn't need rescuing. Odds are that corals, an ancient group of invertebrates, will persist far longer than we will, although particular coral reefs that we cherish may not fare so well as temperatures rise and ocean pH drops. Even there, subjectivity clouds preferences. Marine biologists have done work concluding that the diversity and productivity of a coral reef and one dominated by algae and seaweed is quite similar. I prefer the former. If I fight to save the corals, is that saving nature?

There are practical, as well as philosophical, issues. Humans have worked hard to save—as in rescue—pandas, California condors, and other extremely endangered species. But too great an effort at the extinction end of the spectrum of ecological activity can sap energy and resources that might otherwise be applied to maintain the system as a whole.

There are other facets of the environment that we appropriately seek to save, as in conserve or safeguard for the future. Typically these are special places—an extraordinary mountain or river or ancient forest tract or shoreline. Often their specialness derives from the absence of our own species. On a human-dominated planet, that kind of saving—retaining some sense of the wild and untrammeled—actually becomes ever more important.

On a related front, on a much-used planet restoration still has an equally vital role. I live near Little Stony Point, a 25-acre spit that juts from the east bank of the Hudson River a few miles north of the US Military Academy at West Point and just across from Storm King Mountain, which was the focus of one of the great twentieth-century preservationist battles. That fight was over a proposed pump-storage system for efficiently stockpiling excess electric power for when it was needed. It was a smart technology, just in an awful location if one's concern was scenic beauty. And aesthetics, not energy management, was the prime concern of the day.

Stony Point, which was a quarry from the mid-1800s through 1944, is a modest monument to the value of restoration. As recently as 1967, there was a proposal to build a wallboard factory there. But now—thanks to people like Pete Seeger—it is a quiet park, easily reachable by train from New York City, offering an extraordinary view of the Hudson Highlands. The woods on the point are scraggly and vine-ridden, and the topography includes odd serpentine hillocks that were once gravel heaps. But deer roam, bald eagles soar, and bathers swim. The spot shows just how swiftly biota can spring back when people get out of the way. Restoration can work wonders.

Where restoration and preservation fall short now is in shaping strategies for sustaining some mix of species within an ecosystem that we may cherish mainly because that's what we're used to. The field of restoration ecology seems awfully anachronistic in a country where the zebra mussel has utterly upended river ecosystems, where pythons are slithering in the Everglades, where venomous invasive lionfish from Asian waters dominate Atlantic Ocean reefs. No matter how many lionfish barbecues are held, no matter how many pairs of python skin boots get sold using leather from snakes captured in South Florida's annual python roundup, these species are almost assuredly permanent residents.

But, as I proposed at the outset, there is absolutely a grander role for restoration, as well. What needs restoring, to me, is not American nature but "the nature of America." If we make progress on this task, there'll be plenty of room for nature, including the human component, to thrive in ways we can be proud of for centuries to come, even with the many inevitable losses and astonishingly vast changes. And as the rest of the world, sometimes stutteringly, emulates America's environmental achievements, this can all lead to a very hopeful picture of planetary stewardship.

Here's what I mean by the nature of America.

American culture and politics have had extraordinary moments of bipartisan accord on environmental restoration and protection. Consider the overlap between Robert F. Kennedy's 1968 speech on the limits of gross national product as a metric of progress — "it counts the destruction of the redwoods and the loss of our natural wonder in chaotic sprawl" — and Richard M. Nixon's 1970 State of the Union address: "We can no longer afford to consider air and water common property, free to be abused by anyone without regard to the consequences. Instead, we should begin now to treat them as scarce resources, which we are no more free to contaminate than we are free to throw garbage into our neighbor's yard."

That era, in which a Republican (Nixon) created the Environmental Protection Agency, saw broad support for a suite of laws aimed at cutting air and water pollution and threats to rare species. Compare that to today's legislative paralysis on just about any issue, but especially environmental challenges.

How can we restore a bipartisan focus on sustaining a thriving, if human-dominated, environment? First, by getting comfortable with some uncomfortable realities. One is that we are stuck, for better and worse, with a wide range of views on the mix of regulation and freedoms needed to conserve assets that mostly don't fit into the conventional economy — species richness, an equable climate, clean air and water.

In such a situation, a productive strategy is to look for common interests among divergent factions — as with the common interest of many conservatives and liberals in using oil more efficiently (for some, to reduce dependence on foreign powers; for others, to reduce pollution). There's a common interest among snowmobilers and birders in sustaining the great and variegated Adirondack Park — perhaps the perfectly imperfect patchwork model of an Anthropocene ecosystem. There will always be dynamic tension among such factions, but there's plenty of overlap, as well. As a result, overall, the Adirondack Park system works.

In citing the overlapping themes articulated by Robert Kennedy and Richard Nixon, I'm not arguing that we should pursue some wishful halcyon vision of bipartisanship. The University of Colorado environmental studies Professor Roger A. Pielke Jr. likes to describe the tough work of wrangling agreements in such situations by paraphrasing the liberal commentator Walter Lippmann: The goal of politics is not to get everyone to think alike, but instead to get people who think differently to act alike.

There's plenty of evidence that common ground can be delineated and progress made. I once wrote a blog post titled "Energy Agreement Hidden by Climate Disputes" after digesting Yale survey data showing that while Americans, even those with a lot of scientific literacy, are deeply divided over the extent of danger from human-driven global warming, nearly all support smart use of incentives to encourage energy thrift and innovation. Only a tiny, but strident, minority seems to want no policy on energy at all.

As just one other example, there is common ground among libertarians and environmentalists on the need to revisit mandates and subsidies that encouraged massive expansion of corn production, with few, if any, climate benefits given the energy needed to grow the crops and substantial environmental harms from pesticide use in the Midwest and fertilizer runoff choking the Gulf of Mexico.

One way to facilitate the politics of cooperation and compromise is to move past the politics of fighting from the edges. There is plenty to decry when a company or individual despoils the environment. But the "woe is me, shame on you" rhetoric that typified environmental campaigns of the twentieth century is insufficient in attacking issues like global warming, where companies have profited profoundly from extracting coal and oil, but consumers of electricity and gasoline and jet fuel and the like share substantially in the responsibility for the resulting emissions.

Keep in mind that restoring the nature of America is not just about Washington politics and lobbying fights. Perhaps most important of all, there is a crying need, and opportunity, to re-engage Americans, particularly young people, with the non-human world beyond the masking grid of asphalt and glass and glowing LCD screens that hems us in on all sides.

Figure 2. Students who tended brown trout eggs and then fingerlings in a class-room aquarium for several months as part of Trout Unlimited's Trout in the Classroom program release the fish into a stream in the Croton portion of the watershed for New York City's reservoir system. Photo credit: Matt Shih.

The challenge is deeper than fighting the floodtide of digital experience. Biophilia, as so beautifully articulated by E. O. Wilson, seems implicit in us. But it has to compete with the bio-phobia that has long been within us, as well, and has been amplified by modern civilization's preference for the antiseptic and artificial.

An awakening care for the wider living world can come through a wilderness hike or by tending a garden in a repurposed vacant lot. A lasting spark begins to glow each time an inner city student first lifts a log in a park or feels the tug of a fish on a line, each time a drop of pond water slides under a microscope or a dampened acorn sprouts its first root.

Some of the luckiest students I have seen are those who have participated in the Trout in the Classroom program, in which urban and suburban classes tend trout eggs provided by New York State hatcheries, then study and care for the hatchlings for a couple of months and then — in some cases — release the young fish into clear streams in the Catskills that feed into the city's reservoirs.

I saw one such release in 1999:

In a final exercise to remember the fish's Latin name, a class chanted "*Salmo trutta, Salmo trutta, Salmo trutta.*" Then, taking turns with a little hand net, they plunked the fish into the cool, clear water.

The moment those fish flicked their tails and darted into the stream, the consciousness of these students was forever expanded.

Of course *Salmo trutta* is the European brown trout — long a resident of North America, South America, and just about anywhere else anglers have transplanted them.

Welcome to American nature in the Age of Man.

Nature Preservation and Political Power in the Anthropocene

J. R. McNeill

The Anthropocene and Its Precedents

Lately, boatloads of scholars, scientists, and journalists have embraced the concept and term "the Anthropocene," as a shorthand way to recognize the great power that humankind now exerts — clumsily — over some of the earth's basic systems. Both *National Geographic* (March 2011) and the *Economist* (May 27, 2011) devoted stories to it. There are even three new scientific journals dedicated to the concept. This chapter will offer some reflections on the term, its usefulness, and what it might mean for nature preservation if the Anthropocene coincides with decentralization of power in the international system.

The notion that the earth had a history, with identifiable periods, dates back at least to the French naturalist the Comte de Buffon (1707–88). He challenged biblical calculations and suggested the earth was 75,000 years old. He identified seven stages in the earth's history, in the last of which the "power of man supplemented that of Nature." Buffon did not quite imagine the Anthropocene, because he accorded humankind only a supplementary role. A century later, however, Antonio Stoppani (1824–91), an Italian priest, revolutionary, and geologist, did coin a term — *antropozoico* or anthropozoic — to suggest that the modern era under way constituted an age in the history of life dominated by humankind. His neologism did not catch on, but other minds wandered in similar directions.

According to Google N-gram, the word "Anthropocene" first appeared in use in 1958 but disappeared by 1962. Its sustained use dates from about 2002, when aquatic ecologist Eugene Stoermer and atmospheric chemist Paul Crutzen coauthored a piece about it. Since that time, the term and concept have colonized ever larger chunks of intellectual terrain. Naturally enough, there are rival versions of what it means.

The most conspicuous differences concern the beginning of the Anthropocene, for which at least six dates are jockeying for position. Some authors argue for the late Pleistocene, on the grounds that the extinctions of megafauna occurring then were human handiwork and brought large and lasting changes to life on Earth. Others plunk for a date, perhaps around 5,000 BCE, when early agriculturalists had cut enough forest and built enough rice paddies to inject enough carbon dioxide and methane into the atmosphere to stave off the renewal of Ice Age conditions. Two soil scientists maintain the Anthropocene began 2,000 years ago on the evidence of abundant anthropogenic (human-made) soils. A Korean scholar argues for the Little Ice Age. The most common view, however, and the one that Stoermer and Crutzen held, is that the Anthropocene began only with the advent of fossil fuel use and is only as old as industrialization. Some, however, prefer a more general understanding of the Anthropocene, not so tied to fossil fuels, and see it as beginning in the middle of the twentieth century—in effect only as old as Christian Pfister's "1950s syndrome."

My own view, for what it is worth, is that the better choice is a late Anthropocene, beginning either about 1800 or about 1950. Prior to 1800, while human action had many impacts on the earth and the biosphere, and might have had some effect on climate, the rate, scale, and scope of these impacts was modest compared to what came later. A new stage in the history of human impact came with industrialization. Still another arrived in the mid-twentieth century with the advent of tremendous surges in fossil fuel energy use, population growth, urbanization, tropical deforestation, carbon dioxide emissions, sulfur dioxide emissions, stratospheric ozone depletion, freshwater use, irrigation, river regulation, wetlands drainage, aquifer depletion, fertilizer use, toxic chemical releases, species extinctions, fish landings, ocean acidification, and much else besides. So, as I read the evidence, either the Anthropocene began about 1950, or it began around 1800 and entered a phase of acute intensification (sometimes called "The Great Acceleration" in homage to Karl Polanyi) about 1950. But if adherents to one or more of the early Anthropocene versions compile enough evidence, I hope I will have the grace to change my mind.

The Anthropocene and Its Discontents

The Anthropocene idea has attracted critics in its young career. Some critiques are scientific, some political, and some both.

For some geologists, especially some of the tribe known as stratigraphers, the idea of the Anthropocene may serve as a loose term to denote human impact on the environment, but ought not be blithely elevated to the status of a geological epoch. Geological periods, they insist, are marked off by clear signatures in the earth's fossil or rock record, and, they say, there is nothing durable that shows a transition from Holocene to Anthropocene. The radiation layer resulting from the use and testing of nuclear weapons between 1945 and 1964 will apparently linger for only about 100,000 years, and that is not permanent enough to serve as a boundary marker between geological periods.

For some anthropologists, the term "Anthropocene" is unfortunate, a misrepresentation of the modern ecological predicament and an impediment to useful political action. The problem, they believe, derives from the use of "anthropo" signifying humankind when, as they see it, only a small subset of people is truly responsible for the ecological tumult of the industrial era (those people being industrial capitalists). By confusing humankind with capitalists, the term blunts the possibility of action to redress matters, deluding people into supposing that the Anthropocene results from some innate human qualities.

For some conservation biologists, the term is also unfortunate, even if it has scientific merit, because it invites complacency, even despondency, when vigorous action is required to save species and ecosystems from extinction or transformation. If the biosphere is already transformed by human action, if the Anthropocene is here, what is the point of saving parts of it from further human action? If transforming the biosphere is an innate ambition of the human species, as the idea of an early Anthropocene implies, then what hope is there of resisting the human juggernaut?

Nature Preservation in the Anthropocene

If we admit that indeed the Anthropocene has begun, then environmentalists and conservationists will have to confront this thought (they have been confronting the reality for decades). Should they adjust their ambitions, strategies, and rhetoric? Several essays in this book reflect on that theme as it pertains to the United States. Here I will confine myself to one hypothesis concerning the world as a whole and some speculation about its implications.

Whatever environmental and conservation decisions Americans make in the Anthropocene will matter less and less in the world. The onset of the Anthropocene in the nineteenth century coincided with an era of unusual concentration of global political power and cultural influence. From 1815 to 1914 that power and influence was (to a moderate degree) concentrated in Britain. From 1945 onward, the United States enjoyed (and suffered from) a far greater concentration of global power and influence. While most people alive cannot remember any other configuration of world politics, and hence tend to regard American preeminence as normal and natural, it is in fact an extreme departure from the historical norm.

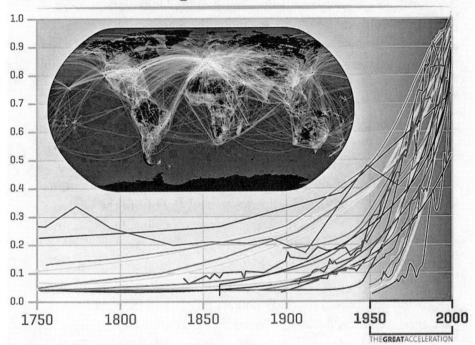

The **Anthropocene** | 24 Indicators, 1 Chart

THEGREATACCELERATION

Figure 3. The International Geosphere-Biosphere Program (IGBP), founded in 1986 to study global change, recently created this chart to encapsulate the confluence of many variables that together show the dramatic human impact on the earth since 1750, and especially since 1950. The selected indicators (i.e., GHG concentrations, population, GDP, domesticated land, species extinctions, damming of rivers, water use) are a haphazard group, but the impression they convey is correct. Whether all this warrants calling the modern age the Anthropocene is another matter, one on which plenty of disagreement remains. Source: IGBP.

For the great majority of human history, political power and cultural influence were decentralized and no state or culture exerted global reach. Only with the transport and communications systems of the nineteenth and twentieth century could any state possibly exert the power and influence that the United States achieved after 1945. But even with those systems intact, and indeed improved, the concentration of global power and influence that characterized the decades after 1945 could not be sustained. It was a moment, a result of the United States being first to turn to oil and to assembly line mass production, and, alone among great powers, emerging from World War II almost unscathed. The United States accounted for about half the world's industrial production in 1945. For a decade or two, despite the Cold War challenges presented by the Soviet Union, the position of the United States remained paramount. But, inevitably, the international political system and world cultural marketplace gradually changed in the direction of normalcy, of decentralization. Beginning in the 1970s, but gathering pace after 2000 or so, power and influence seeped away from the United States, in several directions at once, toward China, India, South Korea, Indonesia, Brazil, and probably on some measures to the European Union. Even as the US economy and military might continued to grow in absolute terms, its relative share of the world economy, the world's military power, and the world's cultural influence, declined. Normalcy—decentralization—began to settle in, and there is every reason to expect this return to normalcy to continue.

As political power and cultural influence flow to the four corners of the earth, the locus of decision making concerning the preservation of nature will migrate too. During the American century (if it can be said to have lasted that long), individuals and institutions in New York and Washington exercised outsized influence, setting conservation agendas and examples. People and governments around the world looked to the US Soil Conservation Service for guidance about holding onto their soil, and to the US Environmental Protection Agency for wisdom about prudent policies on endangered species. But in future no one will have such outsized influence. Decision making will be scattered, localized, as important or almost as important in Shanghai and Beijing, in Mumbai and Delhi, London and Brussels, São Paulo and Brasília as in New York and Washington. This decentralization will shape the world's financial system, its military balance of power, its clothing fashions, and just about everything, including its approaches to nature preservation.

This future means that the United States can continue to set its own agenda for nature preservation but cannot expect many others to follow suit. American preferences for polar bears, pandas, elephants, and whales will carry less weight. American ideas about maximum sustainable yields in fisheries, the appropriateness of using DDT to control mosquito populations, suitable standards for the cleanliness of air and water, and the safety of genetically modified foods will encounter stronger resistance in the international bodies that concern themselves with nature.

Indeed, in the fullness of time, Chinese, Brazilian, Indian, and other cultures will probably exert more and more influence over *American* ideas about nature. If so, then not only will American approaches to preservation become less influential on the global scale, but they will evolve in untraditional directions within the United States as well. American notions of wilderness, always eccentric by world standards, will perhaps regress toward the global mean. Perhaps. An alternative is equally as imaginable, one in which the perception of diminished power and influence induces fierce attachment to national traditions, a sort of cultural fundamentalism. In that scenario, American eccentricities, in approaches to nature as in all things, would become accentuated. And perhaps both will happen at once: the perception of diminished power and influence will inspire some Americans to adopt ideas and practices from overseas, while motivating others to cling tightly to what they regard as genuinely American. In any case, the changing international situation bids fair to revolutionize the nature preservation agenda, its politics, its institutions, and its impacts, first globally, and then, in time, within the United States as well.

Too Big for Nature

Erle C. Ellis

Seven billion people. Two billion more on the way. Intensifying agriculture. Accelerating urbanization. Increasing resource use per person. Atmosphere, climate, and oceans altered by industrial pollution. The ecology of an entire planet transformed by human action.

This is the new normal. We live in the Anthropocene, a new period of Earth's history defined by human influences so profound and pervasive that they are writing a new global record in rock. Humanity has emerged as a global force of nature. The earth will never be the same.

This stark assessment strikes different people in different ways. To some, the idea of humanity playing such a major role in planetary affairs is nothing more than hubris. To others, it marks defeat; as humanity overwhelms the balance of nature societal collapse must surely follow. And to others, the concept is a call to arms — it might still be possible to pull humanity back from the brink and return to harmony with nature. There are other views. Regardless of one's interpretation however, scientific consensus is growing in support of formal recognition of the Anthropocene as a new epoch of geologic time.

We cannot know how long the Anthropocene might last. But implicit in the act of recognizing the Anthropocene is the proposition that it might well endure for thousands of years or longer. Here I approach nature conservation from this Anthropocene perspective, accepting that humanity has already reshaped Earth's ecology and might continue to do so for millennia to come. In so doing, I propose that by embracing the Anthropocene we might enable engagements with nature that yield more desirable outcomes for both humanity and nonhuman nature over the long-term.

The first step in embracing the Anthropocene is to grasp that there is nothing new about human alteration of Earth's ecology. As the most abundant large mammal in history, humans, like other abundant species, have outsized ecological impacts owing merely to our large populations. Yet this fact does not begin to explain how our species came to alter the ecology of an entire planet. The first key to explaining this is that humans are a niche-constructing species. Like the beaver, we engineer ecosystems to sustain our populations. Even more important however, is our species' unrivaled ability to transmit these and other social-technological capabilities across generational time. The human niche has been expanded far beyond anything that unaltered nature could provide, and this has been accomplished through culturally transmitted capabilities that evolve more rapidly than possible by biological evolution.

More than 200,000 years ago, our predecessors used tools of stone and fire to extract more sustenance from landscapes than would ever be possible without these technologies. Our species took this much further. Over generations, our ancestors learned to make use of a far broader spectrum of species after preferred megafauna like the wooly mammoth became rare or extinct, to extract more nutrients from them by cooking and grinding, to burn woodlands to enhance hunting and foraging success, and to propagate the most useful species. Thousands of years before the advent of agriculture, hunter-gatherer societies had already become well established across the earth and depended on increasingly sophisticated social-technological strategies to sustain growing populations in landscapes long ago transformed by their ancestors.

These processes of cultural evolution and population growth

continued to accelerate with the rise of agriculture, urbanization, and industrialization. As toolmakers, burners, propagators, farmers, and urbanites, we have increasingly shaped the ecologies that sustain us. Through millennia of technological innovations, social learning, and ecosystem transformation we have expanded and enhanced the human niche across the planet toward the industrial technologies, urban settlements, and global networks of exchange that now sustain most of humanity. In this continual process of niche construction we became what we are today, the engineers and managers of a planet transformed by the artificial ecosystems required to sustain us. And like our ancestors before us, there is no other way for us to live on this planet.

It is a very good thing that our ancestors developed ever more efficient ways to sustain growing populations on the same old land. We wouldn't be here otherwise. And there is no going back. It would be impossible to sustain seven billion people by hunting and gathering. The same is true even for traditional organic farming. Human populations now depend on advanced technologies like synthetic nitrogen fertilizer that have increased land productivity manyfold over the agricultural systems of even half a century ago. Population growth expected in coming decades will only increase our dependence on advances in technology.

In the Anthropocene, the biggest problem with nature is that we've outgrown it. There is no longer any way to sustain human populations on untransformed ecosystems. To embrace the Anthropocene, we must stop imagining ourselves nurtured by a nonhuman nature and accept the reality that it is only by transforming nature that we survive and thrive. The fate of both humanity and nonhuman nature does not depend on sustaining natural ecosystems but on the most proactive human reshaping of nature ever in history.

How can nature be conserved by such a massive transformation of ecosystems? The answer lies in embracing the role of humanity as permanent shapers and stewards of Earth's ecology. To conserve nature in the Anthropocene, the ecosystems engineered to sustain us must be engaged to the fullest. It is only by increasing the productivity of engineered ecosystems that we gain the ability to leave room for nature. To demand less from our agriculture or our settlements is to demand more from the rest of Earth's ecology. The only hope of conserving any semblance of a wild nature is to offer it the luxury of not serving us.

More than 40 percent of Earth's land already serves humanity directly in the form of agriculture and settlements. Human populations could continue to thrive using just these lands—or even a smaller area. But either outcome will require increasing agricultural productivity and settlement density over the long-term together with more effective sharing of these across society as a whole. While success in this effort is by no means guaranteed and will depend on major sustained economic and social investments, it is certainly within reach. Land can be spared for nature in proportion to how productive engineered ecosystems can be made.

Some large areas still remain unused, especially in the colder and drier regions of the biosphere. It might still be possible to protect their native ecological patterns. Yet climate is changing rapidly. To keep up, species must migrate and colonize new landscapes. Not to allow this will ensure extinction. Nature cannot be locked down. Even in some of the most pristine habitats remaining on Earth, conservation must embrace a changing nature. Anthropocene-aware conservation assists nature in changing—as nature must now change faster than ever.

Sustaining a nature that moves will also require the comprehensive restructuring of the working landscapes that sustain us. Human use of land for agriculture and settlements is rarely complete, leaving fragments of habitat embedded within mosaics of used and unused lands. These remnant, recovering, and less impacted habitats now cover more than one-third of Earth's land and are scattered across Earth's most productive regions. To the extent that these novel habitats and ecosystems can be managed, restored, and connected together to sustain species, they offer perhaps the greatest opportunity of all to sustain biodiversity across the Anthropocene. To make this possible and to facilitate species migration toward the poles, working landscapes must be reengineered at continental scales to offer pathways across the planet for species to move—across and through our fields, our fences, our roads, and our cities. Such work has already begun. And for species too slow to keep up with the Anthropocene, the effort to propagate and transplant them is also well under way.

We must never forget that ecosystem engineering comes naturally to us. Most of humanity transitioned long ago to the hard work of cultivating domesticated species for food. The only significant wild foods remaining in human diets are now harvested from the sea—and with industrial scale technologies, the transition to farmed seafood is moving quickly. While agriculture and industry produce massive global environmental consequences, going back to hunting and gathering or even to traditional technologies would make these massively greater—more land would immediately be brought under the plow. With the giant scale of the human enterprise, to use woodlands for fuel or to absorb our carbon pollution, to use wetlands to purify water, or to demand any other service from an ecosystem is to tangle with the forces of engineering. When we demand that ecosystems service us, we should expect that engineering them could ultimately deliver more. More fast growing trees, engineered wetlands, and restructured landscapes. The result will be a nature shaped more by us, not the other way around.

To conserve any essence of wild nature in the Anthropocene, it will be necessary to consider two natures: one transformed to service us and another that we cannot or will not create or use. To make room for wildness means to engineer the spaces in which we may leave nature alone. By this effort we return to the classic spiritual values of nature conservation—but with a twist. To love nature in the Anthropocene it will be necessary to love an artificial nature, to cherish artificial wilderness along with artificial evolution. Even the ecologies that we work hardest to conserve already feel our touch and are changing fast — in response to our ancestors, cultures, climates, domesticates, weeds, and our machines. To commune now with nature is to become one with all of this.

It is far too late to hold human influences back. Already, our presence is everywhere. To engineer space for nature will demand the most concerted efforts to reduce and remediate the pollution, climate change, exotic species, and other influences that issue from our engineered existence. Yet we can be sure that this cannot ever be accomplished completely. Human influence in the Anthropocene is everywhere and permanent.

In the Anthropocene, humans do not disturb nature. We reshape it. The age-old view of humans as destroyers of nature no longer holds. The Anthropocene demands that we view humanity in the act of creating and sustaining a new nature — and one that will endure in geologic time. Clearly, such a view challenges the proposition that pristine nature still exists and can be conserved. But more importantly, it also challenges the view that by transforming natural ecosystems, we humans are undermining the "life support systems" that sustain us. To those who argue that without ending human transformation of Earth's ecology, humanity must perish along with the rest of nonhuman nature, I offer the opposite. Only to the extent that humanity is able to engineer, design, and conserve nature more actively than ever before will humanity or nonhuman nature thrive in the Anthropocene. The question now is not how nature can continue to sustain humanity, but how humanity can continue to sustain nature.

To embrace the Anthropocene is to become comfortable within the used and crowded planet we have created. In doing so, it is necessary to believe in the possibility that the planet of tomorrow will be no less wondrous to live in than the one we live in today, alive with the species we treasure and also those

Figure 4. The nature that sustains us is a nature reshaped by humanity. Water gathered at the well comes ultimately from nature, but only reaches us through the social-technological efforts of our ancestors. To conserve nature, we must reshape it to sustain us more effectively. Photo credit: Erle Ellis (in Kathmandu, Nepal).

that we don't, and in which the richness of human existence has improved, not declined. Less desirable and even catastrophic futures also exist. It may not be possible for today's human systems and ecologies to be moved by us toward better ends. But let us embrace the challenge to gain mastery over human engagement with the earth. To sustain what has been left to us—the social-ecological legacies of our ancestors—while continuing to discover new ways of living even more desirable than those before and that give us hope, pleasure, sustenance, and freedom in the Anthropocene.

That one species has managed to transform an entire planet is unprecedented. To imagine two billion more of us with even greater living standards than today might seem impossible. The earth is finite, and we have already reshaped most of it. To me this represents the ultimate challenge. We must turn our efforts toward imagining and shaping a future Earth outside of any human experience. In so doing, we must also embrace the reality that much of what we desire must be allowed to emerge by processes beyond our control. Evolution must continue. The nature that we cannot create is growing ever scarcer and more priceless. To create the pristine has always been impossible — and now to preserve it, even more.

After Preservation?

DYNAMIC NATURE IN THE ANTHROPOCENE

Holmes Rolston III

We have entered the first century in 45 million centuries of life on Earth in which one species can jeopardize the planet's future. Since Galileo, Earth seemed a minor planet, lost in the stars. Since Darwin, humans have come late and last on this lonely planet. Today, on our home planet at least, we are putting these once de-centered humans back at the center. This is the Anthropocene epoch, and this high profile discourse comes to showcase the expanding human empire. Humans will manage the planet. We need to figure, perhaps re-figure conservation in this novel future in which we celebrate a new epoch and name it after ourselves.

Preserving and/versus Conserving

There is a widespread distinction, somewhat unfortunate, between nature "preservation" and "conservation." We inherit this from John Muir and Gifford Pinchot. Pinchot, the first head of the US Forest Service, argued his "fight for conservation": "The first duty of the human race is to control the earth it lives upon.... Out of this attack on what nature has given us we have won a kind of prosperity and a kind of civilization and a kind of man that are new in the world." The manifest destiny of Americans is to tame the continent.

Muir thought this arrogant, and founded the Sierra Club, pas-

Figure 5. Anthropocene Earth: control and/or caring?
Source: *Landscape* magazine.

sionately opposing damming (damning!) the Tuolumne River
in California. "Dam Hetch Hetchy! As well dam for water-tanks
the people's cathedrals and churches, for no holier temple has
ever been consecrated by the hearts of man." Woodrow Wilson
authorized the dam in December 1913. Muir lost his last battle.
He died soon after, with — if not of — a broken heart. But we do
have his beloved Yosemite preserved. California now celebrates
John Muir Day on April 21, the first such person honored with
a commemorative day. The state placed him on the California
quarter.

With natural processes, "protect" is perhaps a better word.
Then the question is, Do we want to go "beyond protection"
in this new Anthropocene epoch? But that too comes with a
further question: What is it that we are trying to protect — or
preserve, or conserve — or "save" (to use another word in the
subtitle of this collection)? In more recent vocabulary, what on
Earth are we trying to "sustain"? Sometimes "rescue" is the best
word.

Protecting Products versus Process

Wildland preservation is not museum work. The Starker Leopold report to the national parks in 1963 spoke, regrettably, of
creating "a reasonable illusion of primitive America." This may
be appropriate for the pioneer homesteads in Cades Cove in
Great Smoky Mountains National Park. What wilderness advocates seek to protect today, however, is dynamic and ongoing

wild processes. One can say, if one wishes, that these are prime-val processes. Many natural processes today do not differ much from those of 1492, over half a millennium later.

Baird Callicott, arguing that the wilderness idea should be "revisited," complains that wilderness advocates are "trying to preserve in perpetuity — trying to 'freeze-frame' — the ecological status quo ante" and that this "is as unnatural as it is impossible." He is fighting a straw man. A more sophisticated concept of wilderness preservation aims rather to perpetuate the integrity of evolutionary and ecological processes. Wilderness advocates have never tried to deep-freeze the past; they know processes in constant flux better than most. The day changes from dawn to dusk, seasons pass, plants grow, animals are born, grow up, age, and die. Rivers flow, winds blow, even rocks erode; change is pervasive. Indeed, in wilderness one is most likely to experience geological time. Try a raft trip through the Grand Canyon.

Aldo Leopold, famously, said: "A thing is right when it tends to preserve the integrity, stability, and beauty of the biotic community." He did use the word "preserve," and he evidently refers to processes on the scale of landscape planning where there is relative "stability." Over centuries and millennia, this passes into evolutionary development, of which Leopold was well aware. In the same "Land Ethic" essay, he continues: "Evolution has added layer after layer, link after link." "The trend of evolution is to elaborate and diversify the biota." Such integrity, stability, and beauty is no deep-freeze static balance.

Anthropocene versus Biosphere Conservation

If "Anthropocene" means that humans are dominant on most landscapes around Earth, that has been quite true for centuries. But to recent advocates, the term means more than that: Celebrating what he calls the "Planet of No Return: Human Resilience on an Artificial Earth," Erle Ellis concludes, "Most of all, we must not see the Anthropocene as a crisis, but as the beginning of a new geological epoch ripe with human-directed opportunity." He joins colleagues in the *New York Times*: "The new name is well deserved . . . The Anthropocene does not represent the failure of environmentalism. It is the stage on which a new, more positive and forward-looking environmentalism can be built." The way forward is to embrace an ever-increasing human domination of the landscape, perpetual enlargement of the bounds of human managerial control. Humans are in the driver's seat. The American Geosciences Institute celebrates "humanity's defining moment." Richard Alley provides us with *Earth: The Operator's Manual*. According to Mark Lynas and the *National Geographic*, we are "the God species." "Nature no longer runs the Earth. We do."

The Anthropocene epoch has become a Promethean term, a civilization-challenging idea, an "elevator word," to use Ian Hacking's phrase. Allen Thompson, an environmental philosopher, with a "Radical Hope for Living Well in a Warmer World" (his title), urges us to find a significantly "diminished place for valuing naturalness," replacing it with a new kind of "environmental goodness . . . distinct from nature's autonomy." With "earth systems engineering," Brad Allenby tells us: "The biosphere itself, at levels from the genetic to the landscape, is increasingly a human product." Richard Hobbs and colleagues invite us to envision *Novel Ecosystems: Intervening in the New Ecological World Order* (their title). The *Economist*, in a theme issue, bids us "Welcome to the Anthropocene: A Man-Made World." They foresee "10 billion reasonably rich people" on geo-engineered, genetically synthetic Earth, rebuilt with humans in center focus.

One way of phrasing this question is whether in this artifacted world we wish the Anthropocene to replace the biosphere. Asking about a "safe operating space for humanity," in a feature article in *Nature* in 2009, Johan Rockström argues, using scientific data, that there are nine planetary systems on which humans depend. These can be seen by analysis of chemical pollution; *climate change*; ocean acidification; stratospheric ozone depletion; *biogeochemical nitrogen-phosphorus cycles*; global freshwater use; changing land use; *biodiversity loss*; and atmospheric aerosol loading. Since the Industrial Revolution, in the three systems italicized above the boundaries have already been exceeded. Do we want to preserve/conserve all nine of these systems or to re-engineer them to suit humans better? For at least 10,000 years (what geologists call Holocene times) these systems have remained stable. Surely the wisest course is to keep these major life support systems of Earth in place as they are.

Humans coinhabit Earth with five to ten million other species, and we and they depend on surrounding biotic communities. There are multiple dimensions of naturalness, on both public and private lands. George Peterken, British ecologist, has an eight-point scale. Even on long-settled landscapes there can be natural woodlands, treasured by owners over centuries. There may be native woodlands, often with quite old trees; secondary woodlands with trees 50 to 100 years old; recently restored woodlands; wetlands; moors; hedgerows; and mountains, such as the Alps or the Scottish Cairngorms. Gregory Aplet, a US forest ecologist, distinguishes 12 landscape zones, placed on axes of human "controlled" to autonomously "self-willed" and "pristine" to "novel." Rather than seeking to go "beyond preservation," why not claim that there are and ought to be various degrees of the preservation/conservation/Anthropocene spectrum?

Zoning the landscape, how much human management do we apply where? Which are working landscapes, dedicated to multiple use? This "right-sizing" policy question seems to demand a more specific answer than we actually need to give, if we are concerned (as is this book) with *Saving American Nature in the Age of Humans*. The answer is, Wilderness is the most endangered landscape, the least-sized, the one in shortest supply. Save all you can.

Overall, about 5 percent of the United States is protected — an area about the size of California. But because Alaska contains just over half of America's wilderness, only about 2.7 percent of the contiguous United States is protected—about the area of Minnesota. That too is impressive. But that also means that a little over 97 percent is worked, farmed, grazed, timbered, hunted, dwelt upon, or otherwise human possessed. Surely that is enough, if we have any concern at all for preserving, or conserving, or protecting, or saving—you choose the word— fragments of the plentitude of biodiversity once native to our continent. Surely, if we are to manage more effectively, that ought to be done on the oversized 97 percent we already have taken into our orbit.

Globally, although a diminishing part of the landscape, there are still large areas that are dominantly wild. Yes, human-dominated ecosystems cover more of Earth's land surface than do wild ecosystems. Human agriculture, construction, and mining move more earth than do the natural processes of rock uplift and erosion. But on Earth still today, in an inventory of wilderness remaining, J. Michael McCloskey and Heather Spalding find that all of the settled continents (excluding Europe) are between one-third and one-fourth wilderness. We should have intelligent discussions about how much should remain wild. Confronting such choices, however, let no one say that we have already moved "beyond preservation."

Good for Us/Good of Their Own

Do we save nature because it is good for us, or because the fauna and flora have a good of its own? That vital question has a short answer: Both. The longer answer takes more sophisticated analysis. Becoming educated is becoming civilized. Socrates claimed famously: "The unexamined life is not worth living." He urged: "Know thyself." The classic search in philosophy has been to figure out what is best for us. Environmental philosophers, like myself, claim to be wiser than Socrates: "Life in an unexamined world is not worth living either."

Figure 6. Earth in our hands: caring and/or control? Source: US Postal Service.

Humans are the only species capable of enjoying and advancing the promise of civilization; humans are also the only species capable of enjoying and saving the splendid panorama of life that vitalizes this planet. The totally urban (urbane) life is one-dimensional. To be a three-dimensional person, one needs experience of the urban, and the rural, and the wild. In that sense, the more humans enter the high-tech, artifacted Anthropocene, the more they will be underprivileged. Pushing the 97 percent we inhabit into ever diminishing naturalness is not good for us.

The Anthropocene might prove a dangerous idea, because it impoverishes us. Peter Kareiva and Michelle Marvier, arguing "Conservation for the People" in the *Scientific American,* dismiss the old reason that "we have an ethical obligation to save the world's biodiversity for its own sake." That should be "largely scrapped in favor of an approach that emphasizes saving ecosystems that have value to people." "Human health and well-being should be central to conservation efforts."

But with this focus, we wear a set of blinders; we become blind to nonhuman others. We value our human eyes. Deep sea fish, squid, mantis shrimp, living where light is dim or absent, have evolved spectacular (using the word advisedly) "visual systems that are very different and much more sophisticated than our own," collecting, producing, processing light. "There's a whole language of light down there, and we are barely beginning to understand it," report marine scientists in a news focus in *Science* (March 9, 2012). And so? Out of (our) sight, out of mind? Save them only if they might provide resources for some useful optics or telecommunications research? Or value them for the good of their own supersight, light-fantastic lives.

Every organism has a good-of-its-kind; it defends its own kind as a good kind. Conservation biology did not start with us humans late and lonely; conservation biology has been going on since the origins of life. Such ancient, perennial, and ongoing conservation seems to recognize value in nature as pretty much fact of the matter. Only arrogant humans, ignorant of biology, will claim otherwise.

But this is not simply bad biology. Now it further seems morally offensive for *Homo sapiens*, the sole reflective, moral species, to use its conscience to act only in its collective self-interest toward the rest. Aldo Leopold concluded: "The last word in ignorance is the man who says of an animal or plant, 'What good is it?' If the land mechanism is good as a whole, then every part is good, whether we understand it or not. If the biota, in the course of aeons has built something we like but do not understand, then who but a fool would discard seemingly useless parts? To keep every cog and wheel is the first precaution of intelligent tinkering." Keep it for your tinkering? Or keep it because each of the webworked parts is good in itself and good in the whole? Leopold mixed the two, and so should we.

Maybe some readers will concede: yes, they have a good of their own. But there are bad kinds: rattlesnakes, skunks, malaria germs. Now, however, the burden of proof is on you to say why they are bad kinds, and not just because we don't like them. What we must push for, according to the Royal Society of London, is "sustainable intensification" of reaping the benefits of exploiting Earth. Would not the world's oldest scientific society be as well advised to ask about protecting (*preserving*) ancient and ongoing biodiversity, about whether treading softly is wiser than ever intensifying our imperial exploitation.

If we are to fix the problem in the right place, we must learn to manage ourselves as much as the planet. Be a good citizen, and more. Be a resident on your landscape. True, we must become civilized. True, the future holds advancing technology, but equally: we don't want to live a de-natured life, on a de-natured planet.

Once and Future Nature in the Anthropocene Epoch

We set ourselves toward *Saving American Nature in the Age of Humans*. That seems to be thinking big, on continental and global scales—wise enough in view of present crises. But maybe we need to think further and see a once and future nature. Nature is forever lingering around, even in the Anthropocene. Given a chance, which will come sooner or later, natural forces will flush out human effects. The Holocene covers 12,000 years; there is no prospect of ever-smarter Americans reinventing their continent, or postmodern humans reengineering their Earth for 12 millennia.

A better hope is for a *tapestry* of cultural and natural values, not a *trajectory* even further into the Anthropocene. Preserve/conserve/protect/save/keep nature in symbiosis with humans on this wonderland planet!

Humility in the Anthropocene

Emma Marris

Imagine holding a leatherback turtle egg in your hand. It is round, leathery, the size and shape of a ping-pong ball. Imagine knowing that you had the power to ensure it would hatch and that its inhabitant would grow to its full adult size, a magnificent 1,500 pounds of underwater grace.

One of the founders of conservation biology, Michael Soulé, recently wrote about coming across leatherback turtle eggs being eaten by dogs and vultures on a beach in Trinidad. With his wife's help, he reburied the eggs further up the beach, potentially saving their lives. Soulé admits that the "'rescue' of a few hundred turtle eggs is evolutionarily meaningless." After all, only a few hatchlings make it to adulthood. But he nevertheless found it "thrilling" and personally meaningful. I think most people would. I would love to hold a leatherback turtle egg in my hand and know I was aiding its survival.

But if I care about the turtles, and not just how they make me feel, I won't be saving a few eggs from dogs. I will be doing the kind of work Soulé does when he's not at the beach: scientifically investigating the main threats to the species and addressing them. In the case of the leatherback, that might mean fighting for international bans on longline fishing and drift nets in key turtle migratory areas. Not as thrilling, but more important. Our thrill, after all, is probably less important than actually saving the species.

If you believe that other species are valuable in and of themselves regardless of what humans think of them—that is, if you take a biocentric approach to environmental ethics—then you must prioritize species survival over human values, which include the thrill of helping, the rosy glow of humility or any aspect of our relationship with other species, our relationship with landscapes or even our relationship with wildness. Because it isn't about what I learn from the turtles or how much I care about the turtles. It is about the turtles.

In recent years, conservation has spent considerable time infighting and worrying about its direction. A newer generation of conservationists is shifting their emphasis away from the ideal of wilderness and the creation of new protected areas and toward conservation in all kinds of places. This newer generation is distinctive in its pragmatism, and has been accused by some in the older guard of being irredeemably anthropocentric—caring only about nature in the service of humans. But this is not the case for all of the new voices; many take a biocentric approach, in that they are motivated by the desire to let a rich and teeming nature flourish as well as (or even instead of) by interest in mangroves as storm buffers, forests as resource pools for people, and all the other "ecosystem services" we hear about. So why are they being caricatured as egotistic anthropocentrists? Because they can't shut up about people.

The new wave of conservationists all start with the same old realization: there just isn't that much "untouched wilderness" around anymore. Everything is humanized. Most places once described as "virgin wildernesses" turn out to have been the product of indigenous management by fire, controlled hunting (and sometimes uncontrolled overexploitation), planting, and so on. Indeed, in the twenty-first century, given climate change, there isn't *any* untouched wilderness at all, inside or outside of a park. This point has been made for decades in various fields, but it seems to be hard to really *get*, psychologically.

Thus, in the service of getting it, there is a lot of talk about humanity's influence. Much of this talk has used the concept of the Anthropocene — a proposed geological epoch defined by human activities. Our influence is not limited to turning 75 percent of the earth's surface under the plow or over to our grazing animals. Our cities and roads and radioactive isotopes and synthetic nitrogen will be written into the very rocks. Millions of years from now, we'll be a layer: a layer of Coca-Cola cans and carbon-14 from nuclear testing. We'll be the layer where the mammoth bones stop.

The Anthropocene is one concept people have used to come to terms with humanity's huge influence. Another is the metaphor of gardening. Our vast crop fields and grazing pastures and timber plantations and never-ending series of roads and dams suggest a global-scale garden, where most plants and animals exist where they do because of human desires. I used this metaphor in my 2011 book, *Rambunctious Garden*.

There's a cultural problem with both the Anthropocene and the gardening metaphor. They sound arrogant. People, myself included, running around talking about how humans "control" and "dominate" the planet can sound like assholes. We seem to lack humility.

Here is Soulé, worried that my comparing the earth to a garden means that the new generation of conservationists seek a tame landscape arranged for our gratification: "Will an engineered, garden planet designed to benefit rural and urban communities admit inconvenient, bellicose beasts like lions, elephants, bears, jaguars, wolves, crocodiles and sharks — the keystone species that maintain much of the wild's biodiversity?" (The answer, by the way, is yes! If we manage things sensibly, then ideally we should be able to have more of all of those beasts running free in the future than we do now, even with a larger human population. We may all have to eat vat-grown meat to do it, but I say it is worth it.)

And here's another, from Sandra Postel, head of the Global Water Policy Project: "The term Anthropocene seems likely to expand our hubris rather than inspire the humility we need to pull ourselves back from the brink of planetary transgressions that could well be our undoing."

This instinctual distaste for "the Anthropocene" springs from a conflation of two senses of the idea of "control." In 2014, humans control many of the earth's biogeochemical cycles: we control the climate, the abundances and locations of nutrients like nitrogen, the movement of freshwater, and so on. Here, the word "control" is used in the sense of "influence" or "determine" — as when we say that the weather controls crop yields or insulin levels control blood sugar. It is not being used in the sense of conscious or intelligent control — as when we say that an engineer controls a power plant. We are casually efficacious, as an emergent product of billions of individual actions. But we have no clue what we are doing most of the time.

So when those who talk about the Anthropocene say "humans control the planet," they do not mean humans are *in control*. They just mean that we have become one of the primary drivers of change. And that's indisputably true.

This queasiness over the notion of control brings up a crucial point about the new turn conservation is taking, however. It exposes a potential weakness in a conservation movement without wilderness — a lack of moral purpose. And that's because wilderness is not so much the central physical space for more traditional conservation — indeed, much conservation work has taken place on a very pragmatic basis in all kinds of working lands for 100 years or more. Wilderness is, however, the moral heart of such conservation. Respecting the self-willed character of the land is the moral motivation for many conservationists. Humility before the awesome complexity, great evolutionary age, and sheer beauty of wild nature is the key virtue. A "biocentric" ethic asks us to value the rest of nature in and of itself, and this requires that we accept that we are not the center of the universe or the determiner of value. It requires humility.

Take wilderness out of the picture and whither humility?

Does the earth then become just a pleasure palace or heap of re-sources for humans to arrange according to our desires?

No. I argue that the new emphasis in conservation actually demands just as much from us, morally, as the old. In fact, it asks more. It asks more from us than we are even capable of giving at this stage in our species' life.

A wilderness-centered conservation asks us to withdraw, to get out of the way and let nature run itself. As environmental thinker Paul Kingsnorth put it in an essay in *Orion*, "Withdraw because refusing to help the machine advance—refusing to tighten the ratchet further—is a deeply moral position. With-draw because action is not always more effective than inaction. Withdraw to examine your worldview: the cosmology, the para-digm, the assumptions, the direction of travel. All real change starts with withdrawal." But we can't withdraw without blood on our hands. The climate change we've started, the forests we've sliced into tiny fragments, the plastic we've heaved into the ocean—so much of what we have done has the potential to drive species to extinction if we don't actively intervene to help them.

The ecologist Michael Scott has been making a list of what he terms "conservation reliant species." These are species that need our active assistance to not slide into oblivion—and will con-tinue to need our assistance indefinitely. For these species, the threats to their existence will not go away, and people will have to help them, perhaps forever. For example, saving the endan-gered Hawaiian stilt (*Himantopus mexicanus knudseni*) requires an ongoing program of controlling cats and rats, which have proven so far impossible to eradicate from the islands. The Cali-fornia condor (*Gymnogyps californianus*) only exists in the wild because conservationists periodically release more captive-bred birds. The Karner blue butterfly (*Lycaeides Melissa samuelis*), the red-cockaded woodpecker (*Picoides borealis*), and Kirtland's warbler (*Dendroica kirtlandii*) now rely on us to set the periodic fires on which they depend.

If we withdraw, all these species and many others will wink out.

But Kingsnorth knows this. In the very next paragraph, he urges us to also preserve nonhuman life, perhaps by working for a conservation group or protesting. Traditional conservationists and the new breed of conservationists share the key objective of preventing extinctions. This, to many biocentric thinkers, and to me, is the primary moral duty: stop extinctions.

Traditional conservation also asks us to sacrifice, personally, so that our lives will not pull too hard on the rest of the planet. We must consume less, emit less carbon dioxide, give up beef, and so on. These acts allow us to exercise our virtue. Many of the new breed of conservationists downplay personal sacrifice, and for another pragmatic reason: it doesn't work. Unless the sacrifice is very easy to make or can be made culturally cool, it will not happen for more than a small hard-core population of environmentalists. Yes, we all bring our own bags to the supermarket these days, but most of us still drive there.

New conservationists tend to focus on technologies that make it easier to do the green thing than not. And they tend to focus—perhaps too much, it can be argued—on win-win solutions where companies, developing countries, and nature all come out ahead. This sunny pro-development technophilia seems to jettison humility and sacrifice as part of the work of conservation, making it a creepily amoral exercise.

Paul Wapner, director of the Global Environmental Politics program at American University, worries that the "new environmentalism" is "morally thin." The idea of a world engineered so that "individual, environmental decision-making becomes irrelevant," to him, "raises ethical concerns to the degree that it relieves individuals of having to clarify their moral commitments or take deliberate actions to limit themselves in the service of others' well-being."

Here, Wapner objects to a world in which doing the right thing is too easy because it limits the scope for moral development. But by insisting that a green future for Earth be difficult, old-school conservationists make that green future *all about us*, about our virtue. A truly humble response would be to realize that the green future isn't all about us or our character. Golden lion tamarins and Anhui elms do not care whether we are good people. The rest of nature is indifferent to our moral development. Arranging a conservation program for maximum probability of success, regardless of the effect on human culture, is truly putting other species ahead of ourselves.

If we care about other species, we have to stop worrying about ourselves and whether we are good people and just get on with the work that we are ethically obligated to do. And we have to stop worrying about our cultural creations: our wilderness, our categories of native species, and our senses of place. Those are human goals, anthropocentric values.

Right now, we cannot just put up a fence around a wild place, then venerate it from afar via David Attenborough documentaries. We have to do whatever it takes to keep ecosystems robust in the face of the anthropogenic assaults of climate change, land-use change, pollution, and species movements. And if that means that ecosystem isn't going to be as pretty, or as wild, or won't hew to some historical baseline that is important to us, then so be it.* Aldabran giant tortoises don't care if they are dispersing seeds in the Seychelles where they are native or on Ile aux Aigrettes off Mauritius, where they are playing the role of an extinct tortoise species. Pikas don't care if they are in their native range or surviving climate change on a new peak further north. Lots of endangered plants are happy as pigs in cities. So

---

* Note that in many, if not most, places, the most effective strategy will be "hands off"—but doing nothing in a warming world also counts as a kind of intervention. These unmanaged places will preserve wildness but will be affected by global anthropogenic forces; thus the least managed places may soon be the most novel and unfamiliar. Wildness is still important, but it is no longer synonymous with "primeval" or unchanged.

Figure 7. Effective conservation isn't always pretty. In this case, to protect native marsupials from introduced foxes, cats, and rabbits, a huge fence was erected around a preserve near Mildura in Australia. No one wants to turn preserves into prisons, but we must do what we can to preserve species, even if it means building things and slicing up natural places. Photo credit: Emma Marris.

yes, we can and should protect their native habitat. But also—why not?—plant them in T.J.Maxx parking lots.

Intervening aggressively like this is dangerous. We don't know as much as we would like about how ecosystems work. We can't always predict when our meddling will save species or when it will backfire. But we had better try. We've pulled a few species back from the brink—the California condor, the whooping crane—by insinuating ourselves in their lives as puppet mothers and migration guides, so intimately that I squirm at their lost dignity and wildness. But then I remind myself: that dignity trip is my baggage, not theirs. They just want to live.

Perhaps, through trying, through intervening, we'll learn more and become more effective at "managing" the earth. And that increased ability to consciously control, rather than just blunderingly influence, may well be distasteful to many. They would rather be mere passengers on the earth, taking our place among the other animals, living as part of an ecosystem but not as its master. Me too. That sounds much better all around. Less stressful. More pleasant. But that would be abdicating our responsibility to the many species and ecosystems we've harmed with our lack of mastery. We owe it to them to improve our scientific understanding, our gardening prowess, so we can ensure their continued persistence into the future.

Call it stewardship, call it gardening, call it what you will. It is our job — one I hope we can do all together, with joy — and it is more important than being alone in the wilderness or communing with God in the Grand Canyon or even seeing the condor soar above us. All that is *our relationship with* the rest of nature. To be truly humble is to put other species first, and our relationship with them second. We must not be too proud of our humility that we come to value it above other species. A truly biocentric ethic puts the sea turtle's existence above the condition of the human soul.

The Anthropocene and Ozymandias

Dave Foreman

Much has been made lately of the so-called Anthropocene—
the idea that *Homo sapiens* has so taken over and modified Earth
that we need a new name for our geological age instead of the
outmoded Holocene. One remorseless Anthropoceniac writes,
"Nature is gone...You are living on a used planet. If this bothers
you, get over it. We now live in the Anthropocene—a geological
era in which Earth's atmosphere, lithosphere, and biosphere are
shaped primarily by human forces."

One of the reasons given today for renaming the Anthropocene is that we have so impacted all ecosystems on Earth that
there is no "wilderness" left. Insofar as I know, other than bab-
bling about "pristine," "untouched," and so forth, none of the Anthropoceniacs ever define what they mean by wilderness, which
is not surprising in that none of them give a hint for having been
in a wilderness area or having studied the citizen wilderness
preservation movement.

Moreover, they behave as though their claim about wilderness being snuffed is a new insight of their own. In truth, we
wilderness conservationists have been speaking out how *Homo
sapiens* has been wrecking wilderness worldwide for a hundred years. Bob Marshall, a founder of the Wilderness Society,
warned 80 years ago that the last wilderness of the Rocky Mountains was "disappearing like a snowbank on a south-facing slope
on a warm June day." Congress said in the 1964 Wilderness Act
that the country had to act then due to "increasing population,

accompanied by expanding settlement and growing mechanization" or we would leave no lands in a natural condition for future generations. My book, *Rewilding North America*, documents in gut-wrenching detail how Man has been wreaking a mass extinction for the last 50,000 years or so.

Anthropoceniacs do not seem to understand that when we wilderness conservationists talk about wilderness areas we are not playing a mind-game of believing that these are *pristine* landscapes where the hand of Man has never set foot. Although wilderness holds one end of the human-impact spectrum, it is not a single point but rather a sweep of mostly wild landscapes. Over 70 years ago, Aldo Leopold, the father of the wilderness area idea, wisely wrote that "in any practical program, the unit areas to be preserved must vary greatly in size and in *degree of wildness*" (emphasis added). Senator Frank Church of Idaho was the bill's floor manager in 1964 when the Wilderness Act became law. He understood as well as anyone what Congress meant with the wording of the act. Ten years later, in the heated fight for wilderness areas in the eastern national forests, when the forest service "would have us believe that no lands ever subject to past human impact can qualify as wilderness, now or ever," Church said, "Nothing could be more contrary to the meaning and intent of the Wilderness Act." The words "pristine" and "purity" are not found in the Wilderness Act, which is the best short explanation of wilderness. It seems that intellectual wilderness naysayers, whether wilderness deconstructionists or Anthropoceniacs, if they look at the Wilderness Act at all, see only the ideal definition of wilderness:

> A wilderness, in contrast with those areas where man and his works dominate the landscape, is hereby recognized as an area where the earth and its community of life are untrammeled by man, where man himself is a visitor who does not remain.

In truth, the Wilderness Act has four definitions of wilderness. The first, which I have already quoted, says why we need to protect wilderness. The second, also quoted above, is the ideal; while the third, immediately following the ideal, is the practical:

> An area of wilderness is further defined to mean in this Act an area of undeveloped Federal land retaining its primeval character and influence, without *permanent* improvements or human habitation, which is protected and managed so as to preserve its natural conditions and which (1) *generally appears* to have been *affected primarily* by the forces of nature, with the imprint of man's work *substantially unnoticeable*. [Qualifying words in italics — emphasis added.]

The wish of the Wilderness Society's Howard Zahniser, the main author of the Wilderness Act, and its congressional champions was to keep the idea of wilderness a bit fuzzy. The fourth definition, however, is not fuzzy. It has the lawfully binding language on how federal agencies are to protect and steward the wilderness areas under their hand:

> Except as specifically provided for in this Act, and subject to existing private rights, there shall be no commercial enterprise and no permanent road within any wilderness area designated by this Act and except as necessary to meet minimum requirements for the administration of the area for the purposes of this Act (including measures required in emergencies involving the health and safety of persons within the area), there shall be no temporary road, no use of motor vehicles, motorized equipment or motorboats, no landing of aircraft, no other form of mechanical transport, and no structure or installation within any such area.

Too often, there is confusion between the loose, fuzzy entry criteria for wilderness areas and the tougher rules for management after designation. Before being designated as wilderness, a landscape might have a few roads or acres that were once logged. After designation, however, the roads must be closed, vehicles banned, and future logging prohibited.

So. In the sense of the US Wilderness Act (with over 700 areas totaling over 109 million acres) and like wilderness systems in other lands worldwide, there is, indeed, wilderness. Moreover, some 25 percent of Earth's land is lightly or seldom touched by Man.

But the Anthropoceniacs are really saying that there is no wilderness in its ideal pristine meaning. To answer this assertion, I think we need to put *Homo sapiens* in better perspective.

Life first wriggled on Earth some 3.5 billion years ago. That is a long time. So, let's take an easier timeline and only go back to the unfolding of complex animal life—the Cambrian explosion of 545 million years ago. Make that a book of 545 pages with each page being one million years. With 250 words per page, a word would be 4,000 years.

Where are we? Well, if the last sentence on the last page of the book is a long one of some 13 to 15 words, we behaviorally modern *Homo sapiens* left Africa at the beginning of that sentence. We began to ransack biodiversity then as well. As we spread, we killed the biggest wildlife as we came into new lands. In the middle of the third-to-last word, some of our kind began farming—remaking ecosystems to suit us. In the middle of the second-to-last word, civilizations began.

The very last word in this book of 545 pages takes in the time from 2000 BCE to today. Nearly the whole world met the strictest definition of wilderness until well into the last sentence. Through almost all of that last sentence the share of Earth's biomass held in our bodies grew very slowly. Much of Earth was untrodden by us for thousands of years. Other than the "Overkill" of the "Big Hairies," the wounds we inflicted on the Tree of Life only slowly grew. Not until the last 100 years with our exploding population and systemic pollution of Earth with radioactive fall-

Figure 8. Musk ox bull showing threatening behavior in Noatak River, Alaska, in the heart of an unbroken 13-million-acre wilderness area in the western Brooks Range. This photo was taken seconds before the bull began chasing me. After proving he was dominant, he let me go. One of my friends said, "I can't think of a better place or better way for you to die, Dave, than to be stomped to death by a musk ox here." Photo credit: Dave Foreman.

out, antibiotics, artificial biocides, and greenhouse gases have we finally gotten to the day where we are having an impact everywhere. That is an *impact*, not total control, not the domination of all lands and seas by Man. When I was nearly run down and stomped by a woolly bully of a musk ox bull in a 16-million-acre wilderness area in Alaska a few years ago, I swear to you that Man did not dominate that landscape.

Call the last 100 years the period at the end of the last sentence on the last page of the book of the history of complex animal life. Do you now have a feeling for how long the Tree of Life and wilderness have been without any harm from a ground ape self-named *sapiens*?

I've taken this twisty path to get to my main damnation of the Anthropoceniacs. Though one can hammer them for major mistakes in history and science as many of my friends have done, my beef is with their view of Man's place in evolution and on Earth. It is the *ethics* of the Anthropoceniacs that gives me shudders.

My anger with the Anthropoceniacs is not that they see how Man has taken over Earth (though they overstate this greatly). The first third of my *Rewilding North America* tallies and weighs the ecological wounds we've wrought over the last 50,000 years. I know our impact is great—but not thoroughgoing. By and large, the Anthropoceniacs grossly overstate the degree to which we "control" Earth.

No, my wrath is for the outlook many Anthropoceniacs have toward the ghastly, grisly slaughter of so many wild things. Where is the grief? Where is the shame? Where is the passion to save what's left? Where is the outrage? Where is the sadness for the loss of so many of our neighbors?

Instead, I see many making merry over the coming of the Anthropocene. "We've done it!" they seem to say while high-fiving one another. "Man has finally taken over!" In the writings I've read, they seem blissful, even gleeful. "Now we are gods!"

The mass extinction of other Earthlings seems not to bring them a tear. Witness the words of Peter Kareiva, the chief scientist for the Nature Conservancy, "In many circumstances, the demise of formerly abundant species can be inconsequential to ecosystem function. . . . The passenger pigeon, once so abundant that its flocks darkened the sky, went extinct, along with countless other species from the Steller's sea cow to the dodo, with no catastrophic or even measurable effects." Field biologists and others have shown that this claim is so much biological balderdash—there have been big upsets. However, the true harm, the wound, the loss, the *sin* was the extinction of the passenger pigeon and the ongoing extinctions of countless other Earthlings who have just as much right to their evolutionary adventure as we have to ours. Maybe more, because they are not screwing up things for others. To say the "passenger pigeon . . . went extinct" is akin to a mass murderer saying his victims "became dead." The passenger pigeon did not go extinct; we slaughtered them in a spree of giddy gore in little more than a score of years!

How can anyone who works for something called the Nature Conservancy not feel woe and emptiness at the extinction of the passenger pigeon and all those other extinctions we've wrought and are causing today and tomorrow to make way for our Brave New World—or is it our Brave New Conservation?

Such uncaring, careless, carefree brushing away of all other Earthlings but for the ecosystem services they give the last surviving ground ape is—how can I say this—wicked. It is washed in sin, it is treason to life, to Earth, and to all other Earthlings.

Such Anthropoceniacs behave like our takeover of the Tree of Life was foreordained, that evolution *meant* us and meant us to take over. This is teleology if not theology, my friends, one of the deep misunderstandings Darwin cast out 150 years ago. My children's tale of the 545-page Book of Life shows how we are but one of countless species that come and go. The late Stephen Jay Gould was unsparing on this conceit:

> The worst and most harmful of all our conventional mistakes about the history of our planet [is] the arrogant notion that evolution has a predictable direction leading toward human life.

Man is not the unerring outcome or endpoint of hundreds of millions of years of life's descent with modification, but is, rather, a happy or unhappy (hinging on what kind of Earthling you are) happenstance. We were not "meant to be." Nor is anything Man has done in its flicker of time been meant to be. We happened to become, just as did the curve-billed thrasher getting a drink right now from the birdbath outside my window.

We only happened to be.

This is maybe the hardest lesson from evolution to swallow—one that is stuck in many an Anthropoceniac throat.

It is *Homo sapiens'* arrogance that blinds us to our fate. We think that we, unlike every other species, will live forever. It's not a Thousand-Year Reich we celebrate but an eternal Kingdom of Man Triumphant, of Man over all (*über alles*) other Earthlings. It is we and we alone who decide who lives and who dies, who offers ecosystem services and therefore gets to stay, and who is mere waste biomass. Some may soothe their conscience by making believe this bloodbath, like us, was meant to be. But it is not so. It is our choice to strip off one-third of the limbs of the Tree of Life. We do it willingly, even gleefully, all by our own free will.

The first sentence in Aldo Leopold's *A Sand County Almanac* spells out much of the moral conflict between wilderness and wildlife conservationists and the Anthropoceniacs and their so-called New Conservation (which is truly only the latest version of Gifford Pinchot's resource conservation). Leopold wrote:

> There are some who can live without wild things, and some who cannot.

We who fight for wilderness and all wild Earthlings cannot live without wild things. We believe wild things are good-in-themselves and need offer no services to Man to be of great worth. Those who blithely welcome the Anthropocene and can live without wild things see worth in Nature only in what it offers us as ecosystem services.

The Anthropoceniacs seem to believe that not only is Man running evolution now but that all the lessons scientists have learned about how evolution has worked for billions of years have been thrown out for Man in the Brave New Anthropocene geological era.

One who understood this mindset well, this will to power over Earth, was Percy Bysshe Shelley. Some 200 years ago he wrote:

> I met a traveller from an antique land
> Who said: "Two vast and trunkless legs of stone

Stand in the desert. Near them, on the sand,
Half sunk, a shattered visage lies, whose frown,
And wrinkled lip, and sneer of cold command,
Tell that its sculptor well those passions read,
Which yet survive, stamped on these lifeless things,
The hand that mocked them and the heart that fed.
And on the pedestal these words appear—
'My name is Ozymandias, king of kings:
Look on my works, ye Mighty, and despair!'
Nothing beside remains. Round the decay
Of that colossal wreck, boundless and bare
The lone and level sands stretch far away."

Yes, we can read our tale as the steadily growing sway over Earth by Lord Man. But the Anthropocene technocrats who prattle about grabbing the rudder of evolution and making Earth better are the wanton heirs of a pharaoh's hubris. Their lovely human garden will stand unclothed as either a barnyard or Dr. Frankenstein's lab for other Earthlings. Three-and-a-half billion years of life becomes a short overture before Man in all his Wagnerian glory strides singing onto the set. Does our madness have no end? Have we no humility?

For 6,000 years, each coming age has puffed out its chest. As each Ozymandias falls to the lone and level sands, a greater and more prideful Ozymandias takes his stead. Goodness is overridden more and more by might and the will to power.

Wilderness areas are our meek acknowledgment that we are not gods.

The Higher Altruism

Donald Worster

Only the human species could mourn another creature's extinction or work to protect Earth's ecosystems. It is our unique contribution to conservation. The conservation of energy and matter for the sake of survival are common behaviors throughout the plant and animal kingdoms, but not the conservation of otherness, of wholeness and balance, of endangered communities of life. Those require the evolution of what we might call the higher altruism, an intentional selflessness that may have an element of self-interest but expands to find moral purpose in the act of preservation. Aldo Leopold called it a "land ethic," but we can also call it a more thoughtful and ambitious preservation of diversity, ecological integrity, and wildness on the planet.

America reached a high point of ecological altruism in 1964 with the passage of the Wilderness Act. Like most moral visions, this one was layered over with vestigial language from the past: wilderness as a "resource," wilderness as a place to "use and enjoy," wilderness as an opportunity for "solitude or a primitive and unconfined type of recreation." Those well-worn justifications were the result of more than 60 revisions needed to gain the approval of two houses of Congress, as well as various conservationist groups, who often were still thinking in anthropocentric and utilitarian terms. But unmistakably the act changed the focus of conservation, away from human needs and material demands to the needs of the other-than-human world.

The most telling words in the act were those defining wilder-

ness as "an area where the earth and its community of life are untrammeled by man." To be untrammeled is to experience a freedom of action and self-determination — in this case the freedom
of nonhuman beings from acts of human domination. By this
definition a wilderness cannot be a place of rocks and ice only, or
some lifeless region on the moon. It must be a place where other
forms of life dwell. In those places at least our fellow creatures
are not subjected to our species' economic demands or aesthetic
orderings. Surely the setting aside of such a place primarily is an
act, not of human self-interest, but of humility, selflessness, and
altruism.

Today, the Wilderness Act protects about 110 million acres
in the United States, or 5 percent of the nation's total land area.
A few million more acres are federally owned lands still under
study for inclusion, and still more acres have been preserved
from use by state and local governments and by private landowners. Virtually everything else, more than 90 percent of the
land area of the United States, has been left open to current or
future economic use, and even some of the federally protected
wilderness areas have had, for political reasons, to tolerate pre-
existing mining, hunting, and timber harvesting for a period of
time. So it was only a small step for autonomous nature.

Yet somehow, after five decades have passed, this once small
but popular step in the direction of ecological altruism has become highly controversial. It was not really so in the beginning.
The Wilderness Act admittedly took 10 years to get passed, but
it passed by overwhelming majorities in Congress. Passage came
two years after the publication of Rachel Carson's *Silent Spring*,
at a time of growing bipartisan support for environmental reform, liberal thinking, and generosity on the part of America's
people of plenty. But today it would have a more difficult time
getting through. In fact, it would likely be buried under heaps of
polarized anger and resentment.

The moral cause of preservation remained politically strong
until the presidency of Ronald Reagan, who led a backlash that

tried to brand preservation as a kind of selfishness that would prevent the majority of Americans from improving their standard of living. True, Reagan signed bills adding nearly 11 million acres of protected wilderness. At the same time, however, he appointed to office people who worked relentlessly to open all public lands to oil, gas, and coal development, to tree cutting, mineral extraction, road building, and motorized recreation, who were determined to block the change in moral perspective that wildlands preservation signified. The subsequent rise of neo-conservatism in American society has tended to accept conservation for narrow economic purposes while rejecting conservation for more altruistic ends. The Reagan legacy has often forced preservationists to reemphasize more human-centered goals (e.g., wilderness protection for its tourist potential) and to pursue their more radical goals on private instead of public lands.

More surprisingly, the moral vision of the preservation movement, its commitment to saving and freeing the earth's community of life, has recently come under criticism by critics on the left, who make strange bedfellows with the neo-conservatives. Preservationists, we are now told by a growing number in the "save the humans" party, lack a sense of social justice. They want to protect nature from exploitation not only by capitalist ranchers, oil companies, and real estate developers but also by those who are relatively weaker in terms of power or money, whether they are American Indians or peasant farmers in Africa. Anyone who pursues a preservationist vision stands accused of indifference toward the economic needs of the world's poor. Protecting wilderness and wildlife has become, by this reasoning, an act of aggression against vulnerable people, who want and need to exploit the oil, wood, or game that nature offers. To exclude people from any part of the natural world, it is argued, is to deny those people's rights and to collude in their mistreatment.

Both kinds of anti-preservationist thinking have competing

moral visions of their own to advance, but both share an older, well-entrenched anthropocentrism that is resurgent across the globe. One group wants to set individuals free to pursue their private economic ambitions, to get as rich as they choose, while the other group wants to liberate the bottom ranks of society from deprivation and give them resources they need for self-advancement. Neither group seems willing to set any limits on human exploitation of the nonhuman world. They disagree on whose freedom is most desirable, the rich man's or the poor one's, but not on the goal of removing limits to human appropriation of the earth. In effect, both groups define their vision in terms of "justice" in narrowly human terms, even as they disagree on what justice means.

Fifty years ago it did not seem so necessary for these different moral visions to compete against one another, for the natural world then seemed more open and abundant. That is no longer so. Old-fashioned anthropocentrism is now rising as we face the prospect of a shrinking planet and people become more inclined to fight against a cultural shift toward a biocentric or ecocentric perspective. We no longer are sure of the abundance we once had, and in that mood we are no longer willing to accept the costs that the higher altruism entails.

A pragmatic conservationist like Aldo Leopold, a founding member of the Wilderness Society and architect of the land ethic, would probably respond that all those moral visions are worthy to some extent. All have emerged for good reason, and none is absolutely true or deserves exclusive authority. Leopold would surely grant that social justice is an important principle of modern ethics, even though it is notoriously vague and lacking consensus. So also is the ethos represented by economic individualism; it too has a certain claim to legitimacy, although once again it is loosely defined and resists social or environmental constraints. Likewise, many citizens might grant that the preservation of nature is a worthy ethic but one that lacks clarity and asks too much. Preservation, they might allow, is a legitimate cultural response to a shrinking planet, to an age of global environmental crisis, including mass species extinction and ecosystem destruction. Any pragmatist would ask, why can't we accept all these different visions and work toward an accommodation in which no vision claims absolute truth or demands exclusive importance? But such pragmatism does not come easily to a species that has a long record of fighting over rival ways of thinking.

Right now in the United States the imperative of economic individualism holds the trump hand, while the imperative of social justice has achieved considerable power too. In contrast, the moral vision behind nature conservation (aka preservation) is getting squeezed to the margins. It is younger, more vulnerable, and more radical. Realistically, it cannot expect to become equal to the others any time soon, and no one should expect it ever to replace the others completely. Nor should it do so: an uncompromising or absolutist ecological altruism might lead to a world devoid of all people. Humans, everyone might agree, have a right to appropriate enough resources to live in modest numbers and with modest consumption, and no form of the higher altruism could succeed if it insisted on denying human needs completely. Thus, the question posed by the Wilderness Act and indeed all preservation must be, how far can we or should we go in protecting the autonomy of life beyond people?

Evolutionary biologists have a lot to say about how and why altruism has appeared among humans and other species. They talk about its evolution in purely instinctual terms—a mother giving her life for her child, a single ant dying for the good of its colony. The higher altruism, in contrast, is less clearly instinctual genetic development; it seems more clearly a cultural invention, requiring conscious intention based on a growing knowledge of conditions on Earth. But like older, more biological forms of altruism it requires the sacrifice of self- or group-interest for the good of some greater unit. When we preserve a mountain forest, a lake or ocean, a grassland or wetland, we impose costs on ourselves and on the human community. When a nation acts in this way, as Americans did in 1964, those costs are, or should be, distributed to all citizens. When the international community agrees on preservationist goals, it imposes costs that should be shared across national borders. Who should pay them? Probably no preservationist would argue that those costs should be paid by the poorest and weakest, but rather by those who can most easily bear them — rich people and rich nations.

A careful calculation and distribution of costs across the social ranks and geographical borders of humanity has not always been practiced in the past, whether for the preservation of wildlife, parks, or wilderness. Sometimes only a few have born the costs by sacrificing their basic livelihood. That is not right. Yet it must also be said that it is not always easy to decide how compensation should be calculated. Have the Blackfeet of Montana, for example, been unfairly compensated over the past century for their loss of lands incorporated into Glacier National Park? Have villagers in India been unfairly compensated for yielding their land claims to save the tiger? Have farmers everywhere been unfairly saddled with the costs of preserving habitat for migrating birds? Have mining companies been poorly compensated for giving up legitimate claims to minerals in order to save an endangered ecosystem? These are not easy matters to determine. Unless we do so, the cause of preservation will be resisted. Perhaps we need to establish a better, fairer way to make such determinations — say, an international court charged with distributing fairly the costs of preserving the world's wild places and threatened habitats.

Such a court would have a lot to do, for there is still, even on this rapidly shrinking planet, plenty of nature's autonomy left to preserve. Most of our untrammeled lands and waters lie in the higher latitudes and elevations — in Antarctica or Greenland or Central Asia — and protecting such places would do relatively little toward protecting the world's community of life. It is in the tropics and temperate zones where wild areas become more and more scarce each year and where preservation is especially needed. The challenge of preserving wildness in any of those places, whether life is scarce or abundant there, and fairly distributing the costs among humans would keep an international court busy for at least another century.

Only those who do not accept any part of the moral vision of preservation think there is nothing left to do. They talk of the whole earth as though it were a thoroughly managed environment, a cultural landscape where "pristine" nature no longer exists. This is a reductive and absolutist way of thinking. It is blind and indifferent toward the wildness that still exists, toward the likely event of a peaking and then massive decline in human numbers beginning after the middle of this century, and toward the resilience of organisms to recover and survive. The higher altruism does not require us to follow an impossible standard of Edenic purity. It does require us to care about any and all life that transcends our human boundaries and sympathies.

The Anthropocene

DISTURBING NAME, LIMITED INSIGHT

John A. Vucetich, Michael Paul Nelson,
and Chelsea K. Batavia

Upon seeing a new plant, the first question an amateur bota-
nist asks is, what's its name? That is often also the last question.
When we know a thing's name, we think we know a great deal
about it. When we are sick, we are desperate to know the name
of the disease. Cure or no cure, we receive some comfort knowing that the disease has a name and knowing what that name is.

Any sailor will tell you that renaming a sailboat is not to be
taken lightly. Some suggest three separate ceremonies: one to
remove the previous name, another to de-name the boat, and
still another to rename the boat. Poseidon, the god of the sea, is
said to personally register the name of each and every boat in his
Ledger of the Deep. Callousness or ceremonial miscues are believed to evoke the wrath of the sea god. Penalty can range from
mechanical failure to shipwreck.

The act of naming is serious business.

Disturbing Name

By 2016, the International Commission on Stratigraphy's Working Group on the "Anthropocene" will formally decide whether
or not we live in the Anthropocene, literally the epoch of
humans. Many scholars with no significant knowledge or interest in geology are not waiting for permission to use that name.
They conceptualize the Anthropocene in various ways. Some,
for example, are relatively descriptive, referring to "an unprece-

dented period of profound global change as a result of human activity." Other conceptualizations have an overtly normative overture, characterizing the Anthropocene as "the human centered period on Earth," "a geological epoch defined by our very presence," and even "the Age of Humans."

Whether the conditions are right for demarcating a new geologic epoch on the basis of stratigraphy or other geologic processes is a judgment best left to geologists. However, geologists do sometimes select peculiar names for various segments on the geologic timeline. For example, geologists divide the history of the earth into three periods. The third period is named "Quaternary," meaning "fourth." The history of geologic science explains why the Quaternary has this name—we do not doubt an explanation exists. The current epoch within the Quaternary is named the "Holocene," meaning "entirely recent"—a name that, two or three epochs from now, might seem a bit silly. We are simply pointing out that geologists might not always display the best judgment in naming. This time the consequences of naming are significant.

Our concern with the name "Anthropocene" is the considerable risk it represents for reinforcing and perhaps celebrating a poor relationship between humans and nature. Naming something or someone after oneself runs the risk of great hubris. Hubris is one of the great problems with our relationship to nature. So why would we give a name—to something as grand as a geologic epoch—that risks encouraging or celebrating further hubris?

To some the label "Anthropocene" serves as a reminder that the condition of the world is now harmful to humans. If such a reminder were important, it would be wise to avoid a label risking confusion with a celebration of human dominance, and to choose a more accurate one—such as "Malanthropocene."

Naming the current epoch, the "Anthropocene" or "Malanthropocene" might not motivate anything at all. It may only inspire disempowerment and undermine efforts to heal our relationship with nature because it has been ruined beyond the point of healing. That kind of hopelessness will not serve our desperate need to heal our relationship with nature.

The name "Anthropocene" also runs the risk of indulging misanthropy, the idea that humans are inherently bad for nature. Kathleen Dean Moore reminds us that "we don't name new epochs after the destructive force that ended the epoch that came before." If that were a wise basis for naming, then Moore suggests (with considerable sarcasm) we also consider these alternative names:

> Name the onrushing epoch after a place where the boundary between the rubble of the old era and the new is clearly seen? Then perhaps we are entering the Dubai-cene, for that mirage city built of petroleum... If we name it after the layers of rubble that will pile up during the extinction of most of the plants and animals of the Holocene — the ruined remains of so many of the living beings we grew up with, buried in human waste — then we are entering the Obscene Epoch. It's from the Latin: *ob-* (heap onto) and *-caenum* (filth).

Is and Ought

Many scholars invoke the idea of "our living in the Anthropocene" as an *argument* for why we ought to begin relating to nature in one particular way or another. Some conclude that living in the Anthropocene means we ought to begin living within earthly limits or planetary boundaries. Others conclude we ought to begin geo-engineering the oceans and atmosphere. How can one circumstance give rise to such wildly different conclusions about how we ought to behave?

The problem is that "living in the Anthropocene" is not an *argument*. It cannot, by itself, support any conclusion for how we *ought* to behave. To say that we "live in the Anthropocene" is to describe a circumstance, to pronounce a condition, to depict a certain state of how the world *is*. Because the Anthropocene is conceptualized in such varied ways, it is often not clear precisely what circumstance is being referenced. The problem with that kind of logic is laid bare by one of the most basic principles in ethics: the centuries old idea that *ought* does not, as a principle of logic, follow from *is* alone.

Disregard for the logical necessity of ethics is illustrated by the vision for conservation in the Anthropocene promulgated by Peter Kareiva and Michelle Marvier. Their view begins with a critique of M. E. Soulé's vision for conservation (in the Holocene) whose foundation is a set of explicitly normative premises. Those ethical premises are (1) diversity of organisms is good; (2) ecological complexity is good; (3) evolution is good; and (4) biotic diversity has intrinsic value, irrespective of its instrumental or utilitarian value. These ethical premises may be appropriate (or not), they may be sufficient (or not), and they may have been mishandled by Soulé (or not). The salient point is that Soulé recognized the logical necessity of invoking ethical premises in drawing conclusions about how we ought to behave.

Kareiva and Marvier explicitly dismiss the need to rely on those or any normative principles when they write: "We deviate from this approach and, instead, offer practical statements of what conservation should do in order to succeed." Those "practical statements," however, represent strong support for an anthropocentric ethic and worldview. Anthropocentrism is not a fact that can be deduced exclusively from the premise that we live in a period of profound global change as a result of human activity, or from any descriptive claim about how the world is. Anthropocentrism is an ethical claim, and a deeply contested ethical claim at that, that requires an argument with explicit reference to ethical premises (in addition to claims about how the world is). To see how that argument is not merely inadequate but entirely lacking from the vision of Kareiva and Marvier, one only has to compare their essay with the substantial literature dealing with the ethics of anthropocentrism and nonanthropocentrism.

Persuasive and influential as they may be, similar concerns rise from, for example, the writings of William Steffen, Paul Crutzen, and colleagues in their implied presumption that anthropocentrism is the foundation for our relationship with nature and tacit support for focusing on technological "solutions" without adequately appreciating the problem of overconsumption. We are not saying these authors fail to make any sound and valid arguments. They do. They provide robust arguments for the conclusion that we live in a period of profound global change as a result of human activity. Those arguments are powerful for implying that such a conclusion is tremendously relevant for understanding how we ought to behave. However, many who assert how we ought to behave (in the Anthropocene) do not actually support that assertion with adequate argumentation, and often offer no argumentation at all.

Steffen and colleagues succinctly summarize these concerns when they write:

> The Anthropocene is a reminder that the Holocene, during which complex human societies have developed, has been a stable, accommodating environment and is the only state of the Earth System that we know for sure can support contemporary society.

The subtle but deeply important sin of omission in that sentiment, which seems to permeate their writings, is failing to ask the question, "What aspects of contemporary society ought we continue supporting?" Are hubris, greed, injustice, disregard for the nonhuman world, and overconsumption the elements of contemporary society that we ought to continue supporting?

Limited Insight

Questions about how we ought to relate to nature and what counts as a wise and healthy relationship with nature have always been difficult, weighty questions. For example, Is conservation an anthropocentric endeavor or a nonanthropocentric endeavor? Is an ecosystem healthy to the extent that humans have not affected it? Or is an ecosystem healthy so long as it produces what we want without diminishing its future capacity to produce what we want?

Consider the meaning of sustainability, which might usefully be defined as "meeting human needs in a socially just manner without depriving ecosystems of their health." Depending on how a society understands concepts like ecosystem health, sustainability could mean anything from "exploit as much as desired without infringing on future ability to exploit as much as desired" to "exploit as little as necessary to maintain a meaningful life." Those two attitudes represent wildly different ways of relating to nature and would result in wildly different worlds.

Questions about the goals of conservation and our relationship with nature are difficult to answer. They were difficult questions in the Holocene, and they will be difficult in any new epoch. To simply add "in the Anthropocene" to the end of a question like "What is sustainability?" adds little insight for *how* we should answer the question, and the conceptual obstacles to answering those questions are no more or less weighty.

Recognizing that we live in the Anthropocene (or that we live in a period of profound global change as a result of human activity) certainly constrains the range of options for how we could possibly behave. Those constraints are not always appreciated, though they should be. For example, the existence of seven billion humans obligates us to feed seven billion people, but does not specify how we go about producing the food to do this, nor does it specify what should be done (if anything) about how many people there might be in the future. Nevertheless, within the constraints that exist there is a great deal of latitude for how we might behave, and highlighting that we live in the Anthropocene adds little insight for understanding how we ought to behave, given that range of options.

Perhaps adding "in the Anthropocene" (as in "What is conservation in the Anthropocene?") raises the stakes to the question. Forty-five years ago there were six billion people on the planet; today, more than seven billion. Certainly the stakes are higher—though they have been high for quite a while. Those high stakes in the past did not inspire us to demonstrate any great aptitude for developing broad consensus for wise answers to questions about how we ought to relate to nature. It is far from obvious that our aptitude will improve simply by suggesting the stakes are higher.

On the contrary, the prospect of the Anthropocene has led many to regress to particularly primitive logic. We seem to be developing and condoning a scholarly habit that represents its own new class of logical fallacy, *Argumentum ad Anthropoceneum*. The structure of this invalid argument is

Premise 1: We live in the Anthropocene.
Conclusion: Therefore, we ought to X (for X substitute whatever behavior you like)

Conclusion

"So out of the ground the Lord God formed every animal of the field and every bird of the air, and brought them to the man to see what he would call them; and whatever the man called every living creature, that was its name" (Genesis 2:19). That act of naming has been associated with our despotic relationship with nature. It may pale in comparison to the despotism associated with naming the next geological epoch after ourselves.

Objecting to that concern by insisting that the Anthropocene is simply an objective reality (i.e., living in a human-dominated world) only heightens the concern because that insistence too easily becomes an inappropriate basis for endorsing that despotism. That is, the "Anthropocene" is disturbing in each case that it has been used to promulgate some ethical orientation, but does so under the guise of science. To do so is to misuse two great institutions of civilization — science and ethics. It is a misuse that risks considerable harm to the environment, human welfare, and our humanity.

Robust arguments have already been made for how and why the key to wise relationships with nature depends on a set of virtues that include precaution, humility, empathy, and rationality (i.e., the capacity to articulate a sound and valid argument comprised of premises invoking scientific and ethical principles and the employment of that capacity in decision making). The need to have exercised those virtues was vitally important (and largely neglected) in the Holocene. Those virtues will be vitally important in any new epoch and will indicate the wisdom of, for example, various forms of geo-engineering and the extent to which ecosystem health will depend on human intervention. The deep concern is that we live in a culture with too little capacity or interest in those virtues. Moreover, hubris and misanthropy are serious obstacles to that set of virtues. Naming the Anthropocene seems to work against our need to become familiar with and practiced at those virtues.

Ecology and the Human Future

Bryan Norton

What should we expect if the current rate of transformation of nature into culture continues unabated for 50 or 100 more years? I will not hazard guesses as to the outcome, but in my own expectations I pretty much vacillate between two of the three possibilities: optimism, pessimism, and fatalism. Of these I reject the latter; the stakes are too high to give up, and even if we fail, I demand that we try. But this leaves me swinging between optimism (the apparently naive hope that humans will collectively live together in peace and protect the health of the planet, buoyed by faith that the ability, if not the commitment, is there to save the earth) and pessimism (supported by countless apparently insoluble problems and, for the most part, human indifference toward them).

So I will not predict the future; instead I will try to identify one of the major variables affecting whether future outcomes meet the expectations of the optimists or the pessimists. If humans succeed in creating a livable future on Earth, then they will do so by embodying nature's wisdom in their understanding and actions. One determinant, then, of how humans will live in the future will be how much humans learn from nature as they transform it. Will humans learn from nature? Will they transform their thinking and their valuing to embrace what has recently been called "ecological wisdom?" Or will we obliterate nature, ignore what we can learn from the study of biological and ecological systems, and hurtle into the future without guidance in making mostly natural systems into functioning, mixed and progressively artificial human-and-natural systems? Will biology become more and more synthetic? Or will we continue to revere life as the original miracle, and grow with natural systems not at their expense?

So the variable — or, rather, the nest of variables — I want to focus on in this little speculation regarding the future is the extent to which the coupled human-and-natural systems of the future are based on a respect for ecological systems and the natural history of places. Note the intentional aspect of the variable; the reference to human worldviews and motivations connotes consciously learning from natural systems of the past and present. Can humans transform their worldviews and their values so as to act in accord with nature's wisdom, to the extent we uncover it?

When Nature Goes Away, Where Does It Go?

It is difficult to avoid the expectation that, over the next half-century, the area and the wildness of the world's wild places will be reduced, and more and more systems will be dominated by humans, their activities, and their technologies. These changes will be so pervasive and drastic that our only hope is to learn, and learn quickly, how to live in coupled human-and-natural systems, and to find a way for us to adapt to nature's patterns even as those patterns change. I, and many others, hope that some of the magnificent works of nature and as much "wilderness" as possible can be saved. I have no doubt that the natural places we save to the extent possible will be the gems of an increasingly fragmented, overused landscape, and I have no doubt that they will be laboratories of the humanities no less than the sciences. We should do all in our power to preserve what can be preserved.

But here I begin by thinking through the future of the "built world," and by speculating about the role of ecological ideas as

Figure 9. Comparative tensile strength of spider-web silk and structural steel are visually represented by the difference in length of the two weights. Spider silk provides an analogy that might allow development of stronger materials. Source: Andreas Feininger, *The Anatomy of Nature* (Dover Publications, 1956).

they shape the systems that will be created by human beings. Consider the fascinating and burgeoning field of study and practice called "biomimicry." Advocates and practitioners of this art look to organisms and natural processes as learning devices, as ways to create desired effects more efficiently and more elegantly than presently available human artifices. For example, modeling echolocation in bats has led to a cane for the visually impaired; and biomimetic plastics have been modeled on insect skin.

What we are discussing, without so far having used the words, is learning from *analogies* and *metaphors* drawn from nature. The examples thus far cited show how specific adaptations developed by other species can guide researchers to better ways than they have so far found to bind objects together or to create compounds of amazing strength. We can think of these as "technical analogies"—humans gaining guidance in solving a particular technical problem that has been faced by organisms in their processes of evolving and adapting. There are also examples drawn from social animals that have patterns of behavior that, if observed in humans, would be described as "cooperation" or "joint production," as when bees in a hive communicate regarding sources of nutrients. At a larger scale, grazing livestock have been shown to cause less damage to pasture when herded in ways following patterns of ancient ungulate grazing in the area in question. These more complex relationships can be described as "process learning."

In a sense, all of these examples involve the embodiment of nature within the technologies and institutions of humans. I have no doubt that in this, the epoch of biology, we will see more examples of this. Consider, however, applications of this idea at an even larger scale, a scale at which humans organize their landscapes and their cities; consider the possibility that human infrastructure of the future will be patterned on the principles and practices of ecosystems. If cities dealt with their waste as nature does, they would probably not collect the waste in a centralized process and then neutralize it; nature would find some way to transform waste into useful products at the site of its production.

We can now articulate two possible answers to the puzzling question, where does nature go? If functioning natural processes and wild species are destroyed and replaced with human-designed and -built structures, with human livestock, and with synthetic forms of life, nature will disappear. If we, in our processes of "development," destroy natural species and processes before we can learn the secrets and the arts they exemplify, we will go forward without a "natural compass." This answer to where nature goes, in my view, will be self-defeating to humans and ultimately a tragedy for all of life. But if, on the other hand, we treat natural systems and their denizens with care, and if we study them intently, our actions that change nature can themselves be transformational as the principles and practices of ecosystems can be reborn in the metaphors we use to design and nurture the systems of the future.

The object of this hope has recently been called, by Wei-Ning Xiang in a thoughtful editorial in *Landscape and Urban Planning*, "ecological wisdom." The hopeful position today can take encouragement from this insight: if we can learn to replace natu-

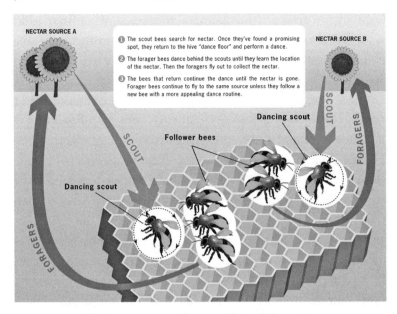

Figure 10. Bees, like Internet servers, face a problem of allocating scarce resources in an unpredictable and constantly changing environment. Bee dances, after visiting a site rich in nectar, provide other bees with competing information about alternative sites, allowing the smooth shifting from one source of honey (analogous to shifting to alternative servers dependent on traffic). Could similar reasoning based on natural processes guide us toward more efficient infrastructure? Source: http://www.newswise.com/articles/bee-strategy-helps-servers-run-more -sweetly. Image used by courtesy of Craig Tovey.

ral systems with human-built and altered systems that respect the wisdom of nature, nature will have been altered, but in ways that continue to express that wisdom. That wisdom has led to adaptive responses to all manner of problems through the history of life; if humans looking to the future develop their habitats according to its wisdom, nature will not disappear but rather will express itself through human actions. Human actions could mimic nature's processes at every level of understanding and being. When nature goes away, it either *really* "goes away" and is obliterated, as when a forest is replaced with a parking lot — or, if humans pursue more thoughtful alterations, nature and its wisdom can migrate into the metaphors, values, and structures that will drive planning and management in the Anthropocene. We can then suggest, at least thinking in extremes, that humanity faces two futures: an Anthropocene that is cut off from its natural roots, floundering into experiment after experiment with no natural guides, or an Anthropocene that is like nature from the beginning of human time in which wisdom involved intimate knowledge of nature.

Metaphors and Mental Models

It is a commonplace to note that the language we use to describe an event or person can make an immense difference, as in the difference exhibited when one person describes an event as an "altercation," while another speaks of the same event as a "vicious attack." Sometimes called the *Rashomon* effect" (in a reference to a film in which the same story is told in totally divergent versions), this widely known phenomenon is the playground of poets and tyrants, and we cannot emphasize its importance enough. Now, couple this phenomenon so pervasive in linguistic communication with another one equally obvious: language changes as society changes. Even as linguistic change enables new conceptions and attitudes, it is difficult to avoid the question, by what process, and with what guidance, does a person *choose* the language and nuance that often expresses their real and deep attitudes? As long as they were "swamps," wet areas were drained and ravaged; today they are "wetlands," and we at least give lip service to protecting them. What happened?

In order to even begin to understand how language and culture evolve together, we must focus on the role of metaphors and analogies, particularly ones drawn from ecology and natural history, in shaping what we say, what we believe, and what we do. It is also important to note that this complex is active at all scales in space and at all levels of consciousness. On the large scale, commentators can say that what humans do to nature may be described as "development," or the same changes can be described as "rape and pillage," or as a tool by which the powerful maintain hegemony over the weaker. On a smaller scale, discovering a new population of an endangered species on a person's property can be described as a disaster or a blessing, depending on the person's values and interests. Such is the plasticity of that most pervasive and elemental tool called "language".

But we should not succumb to *linguistic determinism*, the view that language and conceptualizations ultimately control human actions in a unidirectional manner; in fact, of course, while language has a powerful influence on thought, our activities and cognition also shape our linguistic practices. If we said no more about how language changes, we would fall into a circular conundrum. But we *can* say more about how our language changes — it changes as human actions and cognition are stretched to incorporate new analogies and metaphors, as new *mental models* of our reality are constructed from imagination, experience, and language.

Consider an example. Defenders of Wildlife, an NGO devoted to protecting biodiversity, noted that members of the public showed a surprising indifference to their calls to protect biodiversity. Puzzled, they hired a research team to gather and analyze survey information about the attitudes of citizens toward the natural world. What they found was that when respondents were questioned about their actions and preferences regarding the protection of "biodiversity," they either expressed ignorance or even in many cases showed a negative attitude, as some respondents expressed suspicion that scientists were manipulating them. Surprisingly, however, the predominance of these negative views was strongly reversed when they rephrased the question as one of protecting "the web of life." Once possible concern for natural systems was expressed using this metaphor, respondents showed a positive attitude and a stronger tendency to act to protect the intertwined species and processes that form natural systems. One can speculate that this important difference was driven by the difference between "biodiversity," which emerged as a technical term created by biologists and activists, and the more "natural" understanding that is embodied in everyday interactions with spiders and their handiwork.

So I have identified a key variable that will affect, or at least covary with, the success and failure of societies in addressing the daunting challenges faced by humankind—not to mention the challenges for other species. That variable stands for the degree to which our worldviews are guided by lessons from natural systems. As humans progressively alter what is "natural," it is a justified fear that true nature, the creative engine that experiments ceaselessly with different life strategies—will disappear in the face of the human onslaught. Then the best hope of humankind will be to reinstall nature within the infrastructure and institutions that are developed and applied in planning and management. The way one does that is to explore and apply many metaphors, draw analogies from them, and learn to work with, rather than against, nature.

Metaphors as Guides

Can we say more about the role of metaphor and analogy in human understanding and action? Perhaps. Consider that many of the problems societies face today, to use language introduced by decision theorists, Horst Rittel and Melvin Weber, are *wicked problems*. Rittel and Weber explained these problems by listing 10 characteristics of wicked problems, creating a list with some redundancy and lacking a single theme. These characteristics include the open-endedness of wicked problems, their lack of a definitive solution, and the fact that every problem is unique, meaning chances for learning to do better in subsequent situations are limited. One can explain wicked problems as arising because contestants attempting to solve the problem often do not see the problem in the same way as their counterparts do.

But perhaps we can go deeper than this characterization. I hypothesize that problems are wicked to the extent that different stakeholders hold different *values* affected by a decision at hand: they do not see the problem similarly because they have different interests, hold different positions, advocate different values, and embody different perspectives, and hence they see problems very differently.

Rittel and Webber conclude that when we address wicked problems, we should not expect to achieve a "solution." If participants and stakeholders face different problems, then no unitary solution can be achieved. They are not completely pessimistic, however: we can, through negotiation and compromise, achieve temporary "resolutions"—paths forward that are acceptable to most parties involved. These resolutions can improve some aspects of situations, but the underlying conflicts remain and will reemerge again in response to stresses to the system.

The key point here is that, once we see how varied values and interests create alternative mental models of situations and problems, perhaps we can generalize that, in situations with deep differences about what is valuable in a situation and why, true progress in finding common ground often occurs only when hard-line positions morph into new and perhaps softer-edged differences in dominant metaphors and analogies. The key to opening up shared understandings and common interests, in getting individuals with different perspectives to see things from a common point of view, is triggered by the emergence of a new metaphor.

Looking at wicked problems in this way suggests that, in wicked situations, communication fails because discordant voices advocate based on conflicting values, and we know that those values are embodied in our linguistic forms. Once communities are locked in deep value disagreements, their language tends to become categorical—describing, for example, their opponents as troglodytes or terrorists. Experimentation with new metaphors, through poetry, writing, and theater, can create shared visions sufficient to support cooperation.

So this is the process that gives me hope. The hope I see for the future is that natural metaphors can bridge gaps among conflicting interests, and we can see the common interest in loving and caring for nature. In the biodiversity example cited above, conflict and ignorance stood in the way of any concerted action to protect species, conflict that was exacerbated by the use of a new and unfamiliar term. Once an analogical reference to a well-known natural phenomenon — spiders' webs — is invoked, the goal of protecting biodiversity becomes reasonable to a wider range of the population. This elemental form of learning — and the possibility of new analogical and metaphoric connections it opens up — can make coalitions and collaborators out of discordant groups. Pursuing this path may mean that nature will not go away, but will live forever in the infrastructure of future civilizations.

A Letter to the Editors

IN DEFENSE OF THE RELATIVE WILD

Curt Meine

Dear Ben and Steve,

I appreciate very much the invitation to contribute to your volume, and I apologize for my procrastination. To compound matters, I have been spending way too much time *pondering* my procrastination. But now I know why I have been dawdling. Your title has stopped me in my tracks! *After Preservation.* There it sits, ahead of me on the trail, a patch of quicksand, inconspicuous, covered with leaves, camouflaged so well that I wonder myself if I'm making too much of it. Nothing personal, but my instincts tell me to be wary.

The title implies a premise that I can't accept. It does so subtly, which makes me even warier. It imposes upon the volume, from the get-go, a narrative that I question. It says to the reader (if I may):

Fellow environmentalists: Get with the program. Wake up from your dream of wilderness; the dream, alas, was never anything more than that, an illusion. Conservation, construed as the preservation of wilderness — and it has *always* been construed in that way — has failed. (And by the way, don't think too hard about the evolution of, and differences between, *conservation* and *environmentalism*. That was before your time, and you needn't bother yourself with fine distinctions, or with history. Trust us.) The creation of protected areas was and is the sole technique available to conservationists, and the protection of wilderness from people and modernity has been conservation's overriding goal. That is how we *did* conservation.

But no more. We have learned. And just in time, because it simply does not work in a world of constant change. It most especially doesn't work in a world still in the early stages of accelerating climate disruption, ocean disruption, and ecosystem disruption. The world is all a human artifact now. And that's not such a bad thing. In fact, it can be great for us if we're smart about it. The good news is that, after all, it's a resilient old world. Conservation-as-preservation was merely a phase, a naive one at best, an elitist and imperialist and unjust one at worst, in the development of environmentalism. It is now over, and none too soon. Get over it. It is time to move on. For the first time, we can think about and work on conservation outside protected areas. We can finally begin to care about the rest of the landscape, and the people who are in it.

Perhaps this is unfair and I'm reading too much into two words. But it seems to me that this narrative is not only implied in the title, but is explicit in much of the debate that has played out in the last few years (and in many of the texts that feed it). While I actually concur with many of this narrative's points, I don't believe we can make the debate more fruitful, much less resolve it, as long as we accept the nutshell version of conservation's past that the title conveys. If we consent to a caricatured notion of preservation as a mere historical stage, long dominant, embarrassingly unfit in view of current knowledge and needs, now finally and mercifully being put behind us, we vastly oversimplify the history of conservation, and we perpetuate contemporary challenges in conservation that we are trying to — and must — overcome.

As a way of working myself around the quicksand, let me explore a few side trails.

One trail leads us into the Anthropocene. The presumption is that we have just recently come to understand that we live in a "postwild," absolutely humanized world, and that this undermines the foundations of an environmentalism based on preserving unsullied wild nature apart from people. But the history of the Anthropocene is far more complicated, and more interesting, predating the environmental movement and the conservation movement before it. Steffen et al. (2011) recognize this when they note that "the term Anthropocene may seem a neologism in scientific terminology. However, the idea of an epoch of the natural history of the Earth, driven by humankind, notably 'civilized Man', is not completely new. . . . In retrospect, this line of thought, even before the golden age of Western industrialization and globalization, can be traced back to remarkably prophetic observers and philosophers of Earth history." The authors recognize important historic precedents for this understanding going back ("*before* preservation"?) to the mid-1800s, in the work of such figures as the Italian geologist-priest Antonio Stoppani and proto-conservationist George Perkins Marsh.

To their account I would add another character too much forgotten. In 1883, the Wisconsin geologist T. C. Chamberlin oversaw publication of *Volume I* (though actually the last to be published) of the four-volume *Geology of Wisconsin*. Chamberlin's own studies contributed fundamentally to glaciology, climatology, and planetary geology, and undergird modern understanding of the global carbon cycle and climate change. He himself wrote the first part of the long volume, on general and historical geology. After 15 chapters exhaustively surveying the state of geological knowledge from the formation of the earth to the recent glacial stages—stages that he himself would actually later classify and name—Chamberlin added a short sixteenth chapter, entitled "Psychozoic Era." In a remarkable six paragraphs, Chamberlin suggested that scientists recognize this new era

on a strictly geological basis [italics in original], for it is con-
tended . . . that man is the most important organic agency
yet introduced into geological history. . . . The entire land life
is being revolutionized through man's agency, and to a very
considerable extent, that of the waters. . . . Both the organic
and inorganic agencies of geological progress are powerfully
influenced by him, and . . . a new and profoundly marked
era was inaugurated when he became the dominant organic
being.

Chamberlin apparently coined the term *Psychozoic*, to indicate
that the human influence on the earth "springs from man's intel-
lectuality, more than from his animal force." He concluded his
account in italics: "*This is the geology of the living present.*"

Chamberlin was no obscure scientist. He was among the pre-
eminent geologists of his time, founding editor of the *Journal of
Geology*, president of the University of Wisconsin, creator of the
geology department at the University of Chicago, and president
of the Chicago Academy of Sciences. He contributed significantly to the Progressive-era conservation movement. The fact
that Chamberlin is now so faintly remembered is instructive. Beyond such iconic figures as Ralph Waldo Emerson, Henry David
Thoreau, and John Muir — and beyond our latter-day *perceptions*
(not to mention currently fashionable criticisms) of Emerson,
Thoreau, and Muir — lay a much more complicated background
of evolving understanding of the influence of people on, and in,
nature. There is no doubt that a romanticized "myth of the pristine" influenced the nascent conservation movement. There is
also plenty of evidence that a clear-eyed view of human impacts
on natural systems did as well. And for at least some at the intellectual vanguard, the myth of the pristine was shattered long before the conservation movement identified itself as such.

Here we intersect another side trail. This one involves the
hundred-year record of the American conservation movement,

Figure 11. Male luna moth (*Actius luna*) from a restoration site at the former Badger Army Ammunition Plant in Sauk County, Wisconsin. If we can agree that the "myth of the pristine" was long ago recognized and is now behind us, then perhaps we can agree that the "myth of the humanized" deserves equal scrutiny. And if we can agree on that, then perhaps we may begin to explore the middle ground of the relative wild: the degrees of wildness and human influence in any place, and the ever-changing nature of the relationship between them over time. Photo credit: Curt Meine.

its purported focus on protection and preservation, and its reincarnation as environmentalism beginning in the 1960s. The presumption is that conservation from the outset focused primarily, if not exclusively, on the preservation of "pristine" wilderness, apart from people, in parks and other protected areas; and we are now at pains to exorcise that romantic fixation. This ignores the long history, predating and postdating the emergence of conservation, of stewardship of soils and waters and fisheries and wildlife outside reserves; of conservation on private lands, on indigenous and communal lands, on "working" farms, ranches, and forests, on whole watersheds; of devoted attention to the quality of urban life and care for the urban environment.

Thus we find Theodore Roosevelt not only extolling the Grand Canyon—"Leave it as it is. You cannot improve on it. The ages have been at work on it, and man can only mar it"—and establishing the US Forest Service and the earliest wildlife refuges, but also appointing the National Commission on Country Life, seen in retrospect as "a milestone in American agricultural history... [and] the struggle for sustainability" in the nation's rural landscapes and communities. Thus we hear soil evangelist Hugh Hammond Bennett exhorting farmers, lawmakers, and bureaucrats alike to value soil as the "most fundamental and important of all resources. . . . Shall we not proceed immediately to help the present generation of farmers and to conserve the heritage of posterity?" Thus we witness the creation, in 1935, of the US Soil Conservation Service (now the Natural Resources Conservation Service), dedicated exclusively to conservation, not in protected areas, but on the nation's private lands. Thus we meet Aldo Leopold, his students, and his family in the 1930s and 1940s, pioneering ecological restoration at the University of Wisconsin Arboretum, on Wisconsin watersheds, and at the Leopold farm—"first worn out and then abandoned by our bigger-and-better society"—along the nearby Wisconsin River. Thus we read Lewis Mumford, connecting conservation and the vigor of urban life: "Nothing endures except life: the capacity for birth, growth, and renewal. As life becomes insurgent once more in our civilization, conquering the reckless thrust of barbarism, the culture of cities will be both instrument and goal." Thus we encounter Rachel Carson, triggering the modern environmental movement not by focusing on protected areas, but by opening eyes to changes occurring in the everyday landscapes of farms, suburbs, and cities. There is no doubt that the history of conservation includes a bright and prominent thread of preservation. There is also abundant evidence, for those who care to look for it, that that thread was *never* isolated—that all along it has been woven into the much more complex (and even more colorful) fabric of conservation.

So how did we lose those other threads? One answer (in the United States at least) is that modern environmentalism, as largely an urban and suburban movement emerging in the 1960s and 1970s, left the agrarian and urban conservation traditions behind—even as rural agricultural America became increasingly mechanized, depopulated, corporatized, monetized, and commodified; and as urban America ignored connections to the larger landscapes in which they are embedded. Now, a generation later, younger environmentalists (as well as environmentalism's critics and its reformers) are likely never to know the full richness of the conservation tradition, never to have heard of Liberty Hyde Bailey or Bennett or Leopold or Mumford or perhaps even Teddy Roosevelt.

Another answer is that we didn't really lose those threads after all. They have been carried forward through the neo-agrarian writings of Wendell Berry, Wes Jackson, and Gary Paul Nabhan (among others); through the growth of the local food and watershed and land trust movements (among others); through the now widespread practice of ecological restoration; through robust movements in urban conservation, sustainability, and agriculture. These are not radical departures from an exclusively wilderness-focused environmentalism, but expressions of reclaimed *continuity* with conservation's past, and of connections within contemporary conservation, that point decidedly to the future.

Which brings me to another side trail, one that winds through wildlands and "working" lands and urban lands—and connects them. The presumption is that conservationists and environmentalists have not only been fixated on wilderness, but only pure, pristine wilderness at that; that "traditional" conservation finds value only in absolute wilderness. Of many possible retorts, in 1925, Aldo Leopold, fresh from having successfully advocated for designation of the first "wilderness area" in the United States, on the Gila National Forest in New Mexico, wrote: "Wilderness exists *in all degrees* . . . . Wilderness is *a relative condition.* As a form of land use it cannot be a rigid entity of unchanging content, exclusive of all other forms. On the contrary, it must be a flexible thing, accommodating itself to other forms *and blending with them*" (emphases added). Leopold had no trouble envisioning and acting for conservation across the landscape, advocating forcefully for wildland protection, for conservation farming and ranching and forestry, for ecological restoration, for vibrant, livable towns and cities. He bound them together by exploring *land health* — defined as "the capacity for self-renewal" in the land as an entire, functioning community — as a unifying concept and theme. As an ecologist and a conservationist, he was no purist and he did not segregate wilderness and people. "The weeds in a city lot," he held, "convey the same lesson as the redwoods."

If we can agree that the "myth of the pristine" was long ago recognized and is now behind us, then perhaps we can agree that the "myth of the humanized" deserves equal critical scrutiny. And if we can agree on that, then perhaps we may begin to explore the middle ground of *the relative wild*: the degrees of wildness and human influence in any place, and the ever-changing nature of the relationship between them over time. We can see that reverence for the wild was (and is) not exclusively reserved for wealthy, elite, romantic, Caucasian, Western colonists and imperialists; that such reverence is an inherent aspect of our shared humanity, surfacing in indigenous and agrarian and industrial and postmodern urban cultures alike. We can recognize the wildness in our midst: in our landscapes, in our backyards, on our skin, in our guts, in our souls.

Between the extremes of the thoroughly wild and the thoroughly humanized — and have there been any absolutes since our first ancient hominid ancestor achieved self-consciousness, drew in her first breath, and then exhaled the first molecules of humanized carbon dioxide? — there is a continuum of land-use intensity, from the wild to the urban, bounded above by the atmospheric commons, and below by the oceanic commons. Biological diversity, ecosystem function, and human agency are at work dynamically at every point on and across that continuum. If we seek sustainability and resilience *in the continuum as a whole*, then we may at least make greater common cause. And unless the whole continuum becomes sustainable and resilient, no point or part within it can be so.

Similarly, we can explore the relative wild not just in *space*, but over *time*: human impacts on ecosystems are not uniform or consistent or persistent over time. Human agency intensifies and relaxes, expands and contracts. We now recognize more fully the profound influence of Pre-Colombian cultures on the life and landscapes of the Americas. But now we may move beyond a single, simple dividing line of "presettlement" and "post-contact" cultures in the Western hemisphere. We can see 1491 as just one, albeit highly consequential, point of contact, and put that point into full historical context. We can follow the tracks of our common ancestors within and then out of Africa, splaying out across Eurasia, into Australia, over the oceans and onto islands, into and across the Americas. We can see what we humans have left in our long wake. We can calibrate more carefully the waxing and waning, the character and extent, of human ecological impacts on both sides of ever-shifting, emerging, fading, and still new lines of "contact." We can put into clearer perspective the now global threats we collectively face. This may in turn allow us, in the interest of conservation, to dim the sharp boundary line we are prone to see between our human and natural communities—but to appreciate the shades of contrast we can discern there.

I suppose one could identify conservationists and environmentalists who have placed value on and advocated only for "pure" wilderness and strict preservation, and who have been blind to the social justice concerns that such a stance has, at times and places, entailed. Perhaps every polemic must find, or create, its strawmen. But there is much evidence to the contrary: that for as long as there has been a conservation movement (and arguably before that), there have been people dedicated to making connections across our landscapes and among our varied concerns and goals—before, during, and "after preservation." Wendell Berry frames well the upshot. He writes: "The question before us, then, is an extremely difficult one: How do we begin to remake, or to make, a local culture that will preserve our part of the world while we use it?" To *use* and to *preserve* simultaneously? That may well be a paradox. It may also be a fair mission statement for conservation.

So: the realization that we live in an increasingly humanized world is hardly new; conservation has never been exclusively focused on preserving pure wilderness; and when conservationists *have* focused on protecting wildlands, we should not assume they have done so apart from conservation in the rest of the landscape, blinded by a romantic notion of the pristine. Environmentalism may well have distorted and broken the connections to "pre-environmental" conservationists who envisioned, and to some degree built, continuity of purpose across the landscape. At the same time, it must be admitted: the older conservation movement, for various reasons, could not sustain those connections through the transformation to environmentalism, and on to today. The connections were ignored, forgotten, lost. In our efforts to reclaim them, some apparently hold that we are starting from scratch, here, now. It would not be the first time that those of one generation have claimed to discover, and believed that they did discover, a new idea.

That is what I worry many readers will take from your book's title. Forgive me, guys, for being contrary, and for wanting to avoid the quicksand that I see before me. From where I walk, we moved beyond preservation generations ago. In the working world of conservation, many of us spend most of our time in that place. We also revisit *preservation* with regularity, always confronting both illusions and realities, but always discovering

Figure 12. Gangkhar Puensum ("Mountain of the Three Siblings") in Bhutan. Photo credit: George Archibald.

new things about our places and ourselves, always making us, somehow, simultaneously, more wild and more human. Relearning the relevant parts of conservation history, and bringing them forward into the future, might be one of the most valuable contributions of your volume. I sincerely hope so!

All the best,

Curt

P.S. Just as I am completing this letter, my friend and colleague George Archibald of the International Crane Foundation has written an e-mail to me from Bhutan, where he is counting wintering black-necked cranes in the Phobjikha Valley, and attending the annual festival in honor of the cranes at the Gangteng Monastery. He includes a wonderful series of photos: a gaudily multicolored monal pheasant; the elderly monk who feeds the pheasant (and also feeds George); a beaming weaver and her brilliant silk creation (as colorful and iridescent as the pheasant); a hand-framed quotation on a schoolroom wall ("Earth provides enough to satisfy everyman's need, but not his greed"—Mahatma Ghandi). George attaches another photo, of the great mountain Gangkhar Puensum, it long jagged ridge-line of snow rising above unbroken forest, against a deep-blue Bhutan sky. The Bhutanese have chosen to prohibit mountaineering in deference to local spiritual beliefs. George's comment on his photo: "The highest mountain in Bhutan and the highest unclimbed peak in the world (mountain climbing in Bhutan is illegal—the gods live up there)."

When Extinction Is a Virtue

Ben A. Minteer

Historical accounts tell us that the North American sky was once black with passenger pigeons. Given that at the time of European contact the bird numbered in the billions this was probably only a slight exaggeration. Market hunters, however, would eventually see to it that the sky was clear of pigeons by the second half of the nineteenth century. "Martha," the last individual of the species, expired in the Cincinnati Zoo in 1914. Today, she resides in the Smithsonian's National Museum of Natural History, no longer on regular display and so usually out of sight, though not out of mind.

Writers have long elegized this vanished bird. The great conservationist-philosopher Aldo Leopold issued the most poignant tribute in his 1949 book *A Sand County Almanac*: "We grieve," he wrote, "because no living man will see again the on-rushing phalanx of victorious birds, sweeping a path for spring across the March skies, chasing the defeated winter from all the woods and prairies of Wisconsin."

Leopold could not have known that only a handful of decades after he wrote these words we would be on the verge of a scientific revolution in efforts to reverse the death of species. A prominent group of scientists, futurists, and their allies comprising the "de-extinction" movement argues that we no longer have to accept the finality of extinction. Applying techniques such as cloning and genetic engineering they believe that we can and should return lost species like the passenger pigeon to the land-

Figure 13. The passenger pigeon "Martha," last of the species (d. 1914). The impulse to try to reverse extinction is a strong one, especially in cases where our hand in a species' disappearance is so clear. But the de-extinction proposal raises deep moral and cultural questions about our ability to restrain ourselves in the Anthropocene, even as we continue to alter and transform the wilder corners of the earth. Photo credit: Smithsonian Institution Archives.

scape. This is the goal of the Long Now Foundation, which is actively supporting scientific efforts to re-create the lost bird within its "Revive & Restore" project. But it does not stop there. Scientists in Spain are apparently close to cloning the Pyrenean ibex, a mountain goat that took its last breath in 2000. Other species have been targeted, including the Tasmanian tiger and even the woolly mammoth.

They make some persuasive arguments. The most powerful among them appeal to our sense of justice. In historical cases like the passenger pigeon and the Tasmanian tiger, de-extinction is our opportunity to right past wrongs and to atone for our moral failings. Advocates also point to the sense of wonder the revival of extinct species could encourage among the public. Although we will always have passenger pigeons in museums and books, "book-pigeons," Leopold lamented, "cannot dive out of a cloud to make the deer run for cover, or clap their wings in thunderous applause of mast-laden woods." Recovering vanished species that exist today only in the mists of our ecological memory could thus provide us with a bounty of cultural and aesthetic experiences as we witness their slow return to the landscape.

De-extinctionists argue further that the revived species will enhance ecological function and the diversity of ecosystems, and that they will provide important opportunities for scientists to learn new things about the evolution of lost life forms. Such knowledge could in turn be marshaled for an array of benefits, including the innovation of further genetic technologies (which could be used to preserve endangered species in the future) and the development of new medicines derived from revived plant species.

At the same time, the de-extinction proposal raises significant worries. Resuscitated species could create problems in contemporary environments and for native species that have evolved in the absence of the vanished biota. Increased risk of disease transmission and the threat of biological invasion incite familiar concerns about the potential impacts of species introductions into new environments. Some conservationists also express the fear that, given decades of ecological change and human development, the landscape won't be able to support the revived populations. Others fret about the limited genetic diversity of any "de-extinguished" species and the assumption that reviving a genome is the same thing as recovering the behavior and identity of an animal that evolved over millennia. And there is also the particularly distressing concern that such aggressive manipulation of wildlife might actually end up diminishing our desire (and our limited resources) to conserve extant species—and that it would entail harmful interference in the lives of animals.

These are serious objections. But perhaps the most troubling aspect of de-extinction is what it might mean for *us*. Despite its laudable scientific and conservation motives, de-extinction clearly reflects a new kind of Promethean impulse, one that seeks to leverage our boundless cleverness and powerful tools for conservation rather than for human enhancement. But it's a Promethean urge nonetheless, and if we remember the myth of Prometheus, things did not end very well for him.

Leopold was aware of our tendency to let our gadgets get out in front of our ethics. "Our tools," he cautioned in the late 1930s, "are better than we are, and grow faster than we do. They suffice to crack the atom, to command the tides. But, they do not suffice for the oldest task in human history: to live on a piece of land without spoiling it." The real challenge, he argues, is to live more lightly on the land, and to address the deeper moral and cultural forces driving unsustainable and ecologically destructive practices.

Although employing cutting-edge techniques in genetic engineering and synthetic biology to recover lost species may appear to solve a critical environmental problem (biodiversity loss, the decline of ecological function), de-extinction in the end distracts us from Leopold's more fundamental, and still more elusive task: the development of a moral character and an environmental culture that supports a genuine respect for nature, and that recognizes our own ecological limits on the planet. Few things teach us this sort of earthly modesty anymore; un-engineered nature is one of them. Thoreau captured this sentiment in one of his more memorable lines in *Walden*: "We need to witness our own limits transgressed and some life pasturing freely where we never wander."

Thoreau's preservationist dictum probably seems terribly out of fashion today, especially on an increasingly crowded planet of seven billion (and counting). For that reason, a group of prominent environmentalists, ecologists, and conservation scientists now argue that we need to develop a more people-centered, "pragmatic" environmentalism — one that demands greater control and manipulation of nature, not less. Theirs is a vision that celebrates our technological ingenuity in the "Anthropocene," the Age of Humans. Eschewing Thoreau's advice, these new "ecopragmatists" reject what they believe are tired myths of wild nature, and they rebuke pastoral environmentalists for clinging to a cosseted politics of social action that seeks to contain our power instead of unleashing it on the earth.

Indeed, a boisterous humanism seems to have taken hold in many discussions in conservation circles today, an attitude marked by the rejection of biocentric dogma about an untouchable, sovereign nature. On this score, ecopragmatists would seem to immunize themselves against most preservationist criticisms by evoking a version of the Kantian dictum "ought implies can"; that is, it's unreasonable to demand that we protect nature's integrity from human influence if the very notion of integrity is now a phantom, rendered obsolete by the empirical reality of the global human footprint.

Environmentalist-technophile Stewart Brand is one of the leaders of this charge. In his provocative book *Whole Earth Discipline*, which he bills as an "ecopragmatist manifesto," Brand argues for an aggressive environmentalism heavy on technological fixes and interventions, from ecological engineering and the expansion of nuclear power, to genetic manipulation and geo-engineering the climate. This is far from a new line of argument for Brand. As he infamously once wrote in the *Whole Earth Catalog* (and a remark that has itself been resurrected in de-extinction discussions), "We are as gods and might as well get good at it." Not surprisingly, Brand has emerged as one of the more zealous supporters of the movement to resurrect lost species, establishing a new de-extinction initiative ("Revive & Restore") under the auspices of his Long Now Foundation. "Don't mourn," he exhorts us. "Organize."

On one level, these pleas to be more pragmatic in conservation and environmental management are hard to argue with. Shouldn't we want to be more practical (and therefore more effective) in our efforts to achieve desired outcomes, whether these concern species conservation, or wilderness preservation, or the creation of more sustainable communities across the landscape? Of course we should.

The problem isn't the appeal to pragmatism; it's that the new ecopragmatists have seriously misread one of the tradition's most important moral lessons. When self-described environmental pragmatists like Ted Nordhaus and Michael Shellenberger of the progressive Breakthrough Institute write that, "whether we like it or not, humans have become the meaning of the earth," they are expressing a Promethean view, not a pragmatist one. Unlike many of today's ecopragmatists, American pragmatist philosophers John Dewey and William James didn't place humans at the center of the universe. Rather, their philosophy was steeped in a reflective sense of human contingency and, in Dewey's case, a respect for nature as part of responsible human agency on the planet.

Dewey called this "natural piety," by which he meant a sense of our profound dependency on environmental conditions, a reliance entailing an attitude of human modesty rather than superiority toward the natural world. A faithful reading of American pragmatism suggests an ethic of careful adaptation and cautious adjustment, a philosophy of environmental prudence.

It is an ethic that neither rationalizes the human domination of nature nor recklessly seeks to extend our technological prowess. "Humanity is not, as was once thought, the end for which all things were formed," Dewey wrote in his 1927 treatise on political philosophy, *The Public and Its Problems*. "It is but a slight and feeble thing, perhaps an episodic one, in the vast stretch of the universe."

Self-restraint, precaution, fallibility, ecological modesty— these classical pragmatist virtues have been all but lost in the neo-pragmatist revival in environmentalism. Yet Dewey's emphasis on maintaining a respect for nature within a broader philosophy of democratic humanism remains vital to our current environmental challenge. For one thing, it helps us understand how we might thread the needle between the ideological extremes of an antimodernist primitivism and a technologically exuberant Prometheanism. It also corrects the faulty (though common) assumption that to criticize the excesses of human ambition and technological will in nature is to embrace those outmoded preservationist ideas that have hampered the development of a scientifically rational and politically feasible environmentalism for a human-dominated planet.

The value of this pragmatist "middle way" (i.e., a path running between the pure versions of the "nature first!" and "people first!" worldviews) was not lost on Aldo Leopold. Although he regretted the modern worship of technology and the prevailing utilitarian view of societal progress, Leopold recognized that good conservation often required the intensive and pervasive manipulation of nature. He understood, too, that wild species were increasingly enmeshed in the human enterprise. As he wrote in his seminal text, *Game Management* (1933):

> Every head of wild life still alive in this country is already artificialized, in that its existence is conditioned by economic forces ... The hope of the future lies not in curbing the influence of human occupancy—it is already too late for that—

but in creating a better understanding of the extent of that influence and a new ethic for its governance.

Leopold's qualification ("in that") is significant: he was not questioning the philosophical status of "wild" species, only making a practical observation about the extensive influence of human economic and social forces on wildlife. Although he was a pragmatist who understood that the strategy of "pure preservation" in wildlife management was a nonstarter, Leopold nevertheless believed that we could develop a more responsible relationship to other species and to the land, a relationship that required us to adopt a clearer view of the human impact on the environment and fashion, as he put it, "a new ethic for its governance." Read this way, his "land ethic" becomes as much a political expression of willful self-limitation in a technological culture as it is a biocentric veneration of the "integrity, beauty, and stability" of the natural world.

All of this is to say, then, that there is great virtue in keeping extinct species extinct. Meditation on their loss forces us to remember our fallibility and our finitude. We are a wickedly smart species, and occasionally a heroic and even exceptional one. But we are also a species that can become mesmerized by its own power. It would be silly to deny the reality of that power. But we should cherish and protect the capacity of nature, including species no longer with us, to teach us something profound (and something quite old) about the value of collective self-restraint in the Age of Humans.

Despite some good intentions, the attempt to revive extinct species isn't a genuinely pragmatic response to biodiversity loss or to the challenge of conservation on a crowded planet. Nor is it a proper act of ecological contrition. It is yet another example of the refusal to recognize moral and technological boundaries in nature; to, as Thoreau would put it, observe "some life pasturing freely where we never wander." Although it might cut against the progressive aims of science and technology to say it, there can sometimes be real wisdom in taking our foot off the gas, in fighting the impulse to further control and manipulate, or to "fix" nature.

Resisting the Promethean urge to resurrect lost biological forms (and similar calls to engage in other radical manipulations across the planetary scale, e.g., geoengineering) will show that we understand the value of setting some limits on human intervention in environmental systems even as our global influence grows. The recognition of our unique impact on the land, the honest acknowledgment of the size and depth of the human planetary footprint, doesn't require giving up on the view that many species and ecosystems do and should continue to exist free from significant human manipulation.

Drawing this line is necessary if we're serious about keeping alive a meaningful ethic of nature preservation and preserving a rich and diverse ecological mosaic in this century.

It will hopefully also remind us that some of the most ennobling and transformative applications of human power are not found in displays of technological mastery or environmental control. They reside instead in acts of restraint and self-possession, including the creation of moral boundaries in nature that we do not cross, even if we can.

Pleistocene Rewilding and the Future of Biodiversity

Harry W. Greene

Backdrop

In a 2005 *Nature* commentary, 12 of us proposed partially restoring evolutionary and ecological processes lost with the extinction of almost 60 species of large North American vertebrates, about 12,000 years ago. Paul Martin had first floated "Pleistocene rewilding" (PR) in 1969, on the heels of his controversial hypothesis that human overkill caused those extinctions, but over the next four decades his corollary notion of active restitution failed to gain traction with conservationists. Fast forward to 2004, when PhD student Josh Donlan (experienced in eliminating invasive species from islands) and I (long committed to conserving dangerous animals) — both of us friends with Paul — convened a group of experts and visionaries on a New Mexico ranch to discuss potential benefits and costs of PR. A lengthy manuscript spawned by that meeting was accepted if cut to two printed pages, 10 references, and two authors; we deemed that last stipulation unacceptable, and in the end everyone remained on board.

The *Nature* commentary summarized evolutionary and ecological implications of global megafaunal decline, and argued for end-Pleistocene as a less-arbitrary restoration benchmark than 1492. We emphasized that large vertebrates were once widespread and humans helped wipe them out; that they were and still could be important in trophic cascades, nutrient cycling, and other community- and ecosystem-level processes; and that they're declining almost everywhere. Then we explored prospects for PR via exemplars, including lions, cheetahs, elephants, camelids, equids, condors, and tortoises — chosen to span an array in relatedness to what had been lost, as well as in popularity, danger, economic importance, and conservation value. We imagined three stages, with the first, widespread presence of domestic camelids and feral equids, already under way. A second stage would entail small-scale experimental restorations, and the third would be huge Pleistocene parks, such that megafauna could roam widely and economically benefit local stakeholders. We envisioned PR proceeding with explicit goals and fail-safe mechanisms, conducted against a backdrop of expert consultation and public debate.

A year later our *American Naturalist* paper laid out a detailed framework, responded to critiques of the *Nature* commentary, and ended with a challenge, paraphrased thusly: For those who find objections to PR compelling, will you risk extinction of the remaining megafaunal species, should economic, political, and climate changes prove catastrophic within their current ranges? Are you content that your descendants might live in a world devoid of large wild animals? Will you settle for an American fauna that is severely depauperate relative to just 100 centuries ago? We closed by declaring "the obstacles to PR are substantial and the risks are not trivial, [but] we can no longer accept a hands-off approach to wilderness preservation as realistic, defensible, or cost-free. It is time not only to save wild places but to rewild and reinvigorate them."

Together those publications resulted in substantial critical response, as well as brief, for Josh and me, unprecedented notoriety. During the ensuing years, however, the extinction crisis has worsened, attitudes have changed, and my own views have evolved. PR thus provides a basis from which to reflect on how such controversies play out, and for speculating on how their underlying value differences and dynamics might affect conservation. Accordingly, this essay draws on that experience for some tentative conclusions regarding broader issues.

Aftermath

Reactions to PR were immediate and dramatic. Within a few weeks Josh and I fielded about 50 radio and TV interviews. We answered nearly 1,000 messages, ranging from highly favorable through politely critical to hostile or absurd. One correspondent threatened to shoot Josh and his elephants. Another labeled me a "goofball, dipwad, doofus with a scrambled brain, linked to terrorist atrocities like 9-11," then advised I "stick to playing with lab rats, befriending cockroaches, or collecting dust mites." There was positive coverage by USA *Today*, the *Economist*, and commentators as ideologically diverse as CNN's Lou Dobbs and the *New York Times*' Nicholas Kristoff. Seventy percent of 7,000 people polled by MSNBC viewed our proposal favorably, and we received a fair amount of unsolicited praise, ranging from members of the National Academy of Sciences to Texas ranchers. Negative reactions from colleagues likewise spanned scholarly to ill informed and superficial. We were not permitted to respond to several letters in *Nature*, including one our prepublication review saved from a contrived falsehood, one by someone who admitted writing it to add a hot citation to his CV, and others advocating conservation by the NGOs for whom the writers worked. Most bizarrely, we were accused of plagiarism for not acknowledging earlier mention of PR in papers by our own coauthors, references we had been required to delete by *Nature*'s editors. The *American Naturalist* paper dealt with those letters as well as the only detailed published critique, in the process devoting 164 words to a topic (African tourism dollars and conservation) that an opinion piece five years later spent 64 words claiming we hadn't discussed. Subsequently several philosophers weighed in, their coverage ranging from cogent critical analysis (thanks Baird Callicott!) to little more than petulant name-calling. With few exceptions, critics ignored our closing questions.

After scrutinizing published critiques as well as informal feedback, Josh and I identified several key themes. The commonest criticism is encapsulated in "Haven't you heard of exotic rabbits and cane toads in Australia?" to which we'd respond, "Risk assessment is crucially important, but until less than two centuries ago Oz had few placental mammals, let alone lagomorphs, and no toads"—so that situation is hardly comparable to restoring conspecifics or close relatives of extinct North American megafauna. Likewise, the claim that PR is only "slightly less sensationalistic" than Jurassic Park is belied by the 15,000-fold difference in time-to-present, as well as the fact that the closest living relatives of *Velociraptor* are birds and crocodilians, whereas our proposal advocated using conspecifics or very close relatives as proxies for missing lions, elephants, and so forth.

We answered the "BINGO [big NGO] criticism," that PR would divert scarce money from other projects, by noting that conservation dollars aren't necessarily transferable, novel ideas can generate unanticipated funding, and in any case, shouldn't new initiatives be evaluated on their merits, rather than prohibited a priori? We noted that the "killer criticism" is literally and figuratively just that, because it embodies the real reason PR likely never will be widely implemented in North America: "If people won't tolerate wolves and grizzlies, they surely won't put up with lions and elephants!" In other words, NLIMBY, no lions in my backyard—we want dangerous animals somewhere, just not here, and we want Tanzanians and others to pay the local, very human costs of coexistence.

Daunting prospects of living with natural born killers notwithstanding, the three most conceptually provocative criticisms can be summarized as "North America has changed a lot since the Pleistocene," "those aren't the same elephants we lost," and "preserve African animals in Africa." In response, we emphasized that only 12,000 years ago, about twice the lifespan of an individual bristle cone pine, there were several species of mammoths and mastodons in the United States, and that Asian elephants are more closely related to our extinct forms than to extant African species. Consider also that the last known live mammoth was on an island off Siberia less than 4,000 years ago, about the time Hittite king Mursuli was sacking Babylon. North America indeed has changed mightily since end-Pleistocene — as has Yellowstone in the past few decades — but with respect to overall evolutionary potential and ecological complexity, by far the most significant changes in the last 20,000 years came with megafaunal extinction, not much more than half that long ago.

Now recall two recent success stories. Peregrines were restored to North America with a captive-bred amalgam of seven subspecies, four of them Eurasian, and little or no controversy. Historically, the black-footed ferret occurred with its prairie dog prey from Canada to Mexico, but by the late twentieth century was hailed as our rarest mammal, then declared extinct. In 1981, however, a Wyoming rancher's dog brought in a dead *Mustela nigripes*, and captive-bred progeny from that remnant population now have been restored to sites throughout the species' historic geographic range. No one suggested renaming it the Wyoming ferret, nor blocked using assisted migration to enhance the species' prospects for long-term survival — never mind that many aspects of its former distribution have changed over the past few decades or that extant individuals aren't genetically identical to those lost earlier from the southern Great Plains. To the contrary, however, we persist in calling *Panthera leo*, whose Indian range segment reflects a formerly Holarctic lineage, the "African" lion. Our PR group believed that, given those sorts of complexities, reversing precipitous declines in the remaining megafauna demands deeper-time perspectives and global solutions.

Meanwhile, favorably regarded, science-based rewilding is happening elsewhere, encompassing multispecies assemblages of large herbivores in Russia and the Netherlands, bison in

Germany, wild horses in Spain, and giant tortoises on oceanic islands. By contrast, in the United States we maintain more than 200 cheetahs and 100 Asian elephants in profoundly unnatural captive conditions, ostensibly for conservation purposes, but only the bolson tortoise is being restored to its prehistoric range, thanks to media mogul–rancher Ted Turner's Endangered Species Foundation. The Columbian landfall is irrelevant to those other places, however, and tortoises never eat people.

Onward

Despite some bumps along the way and a few aftershocks, running the PR workshop with Josh and cowriting those manuscripts with our colleagues have been among my most satisfying professional endeavors during almost four decades as an academic. We had spirited, intelligent debates with minimal posturing; everyone readily agreed our only graduate student would be listed as first author for the publications. And given financial constraints—Josh and I contributed some foundation monies, participants paid for their own travel—my only substantial regret is that we didn't include a Native American and a philosopher, about which more shortly. As someone with a lifelong affection for reptiles I'd do it all over just for inspiring the bolson tortoise project, let alone to cast a critical light on restoration benchmarks. Withal, given the controversy, it's worth asking whether there might be broader lessons in all this, and I offer the following candidates.

First, judging from reactions to PR, many environmentalists and biologists are naive about the deep history of North America. There's a pervasive, inaccurate sense that while of course our continent has changed radically since the Pleistocene, biotas have remained constant for at least the last 500 years. In fact, given rapid global change, any one overarching restoration benchmark—end-Pleistocene, 1492, 1950, or 2015—will soon be irrelevant. Humans will, instead, to whatever extent we value biodiversity, manage landscapes with diverse functional and aesthetic goals, engaged in what Daniel Janzen calls "wildland gardening." And given current projections, ever more species will occur where they don't now exist, including some invasives and others we purposely relocate. PR, depending as it does on deep historical and global perspectives, might well provide useful insights with respect to the composition of future biotas.

Second, there is widespread antipathy toward dangerous animals, yet the efficacy of traditional measures for conserving them is increasingly doubtful. Signs at the National Zoological Park's cheetah exhibit say "losing their race for survival" and "if current trends continue it will become extinct," not "send money to your favorite NGO and they'll be fine." Today lions are in big trouble almost everywhere except protected areas, contrary to their seemingly secure status only a decade ago, and they're generally doing best in large fenced reserves. Nonetheless, having long worked for rattlesnake appreciation and conservation, I'm not ready to give up on dangerous animals—but it's an uphill struggle, and the onus is on those of us who appreciate them to inspire other folks to do so. We need fresh ideas and some new arrows in our quiver!

Third, we should have had a Native American ecologist and a philosopher at our workshop, because ultimately the fate of megafauna will come down to land-use and ethical norms—the latter a controversial topic explored in Minteer's essay and exemplified, jarringly at times, by others in this volume. Our challenge as nature-loving wildland gardeners thus will be to convince others to care, yet reverently touting Americentric notions of wild, natural, and intrinsic value no longer seems likely to get the job done. I believe, instead, that soon we will confront questions like the following, questions inconceivable when I was a graduate student and starkly discordant with some of the hallowed environmentalist precepts with which I grew up: Would we rather at least have elephants somewhere or have them nowhere? And if we want them somewhere, who will pay for their continued existence? Of course eventually the proboscideans in a reconstituted North American herd would no longer be "Asian," but they could still matter to *global elephant conservation*, as long as someone, somewhere would put up with them.

I'll push this matter of values a bit further with an example bound to sit poorly with those who view cows as everywhere an unambiguous scourge. Old World aurochs, the wild progenitors of domestic cattle, went extinct in 1627, but more than a century earlier Columbus had brought Iberian stock to Florida and Mexico, such that after some 500 years of mostly natural selection we have respectively divergent Cracker and Longhorn lineages. I often visit a Texas ranch inhabited by the latter, as well as by 65 species of amphibians and reptiles, 35 species of mammals, and a rich avifauna. The Longhorns, more so than Herefords and other "improved" breeds, behave like free-living ungulates — they birth without human assistance and predators don't take their calves; cows exhibit a pecking order, and the bull is attentive to his herd. As it also happens, extinct Pleistocene shrub-oxen were about the same size, Plains Indians portrayed the feral Spanish cattle as akin to bison in symbolic power, and Longhorn meat is almost as lean as that of their shaggier brethren. Knowing all that, the ranch feels wilder for the presence of those magnificent animals. I'm happy they're out there, and as an omnivore, I also look forward to a day when organic food stores proclaim, "Keep Texas wild, proud, and healthy: eat Longhorns!"

Finally, we could use a golden rule approach to professionalism, for our individual sakes as well as for the future of biodiversity. More than ever conservation needs new ideas, risky ventures to find out what will work and what won't; biodiversity doesn't benefit from us calling each other stupid as a substitute for rational discourse. So how about we keep our ego scuffles off to the sidelines? How about we actually read each other's publications before critically commenting on them to the media, and even then refrain from hyperbole and misrepresentation?

Figure 14. Some 200 species of vertebrates inhabit this ranch in the Texas Hill Country—a continental-scale ecotone and among the most biologically diverse regions in North America. The Double Helix is also home to Longhorns, a breed shaped by roughly 500 years of selection in the arid Southwest, and their presence enhances its aesthetic values. Photo credit: Harry W. Greene.

Can't we ask, having encountered an eccentric initiative, "If that idea were mine, how would I like it treated?" Otherwise, as Zoo Atlanta's Joe Mendelson said about another conservationist's self-serving pronouncements, "Oh, I see, *this* is more important than the frogs!"

Let's get on with saving rattlesnakes, lions, and elephants!

The Democratic Promise
of Nature Preservation

Mark Fiege

"Each age writes the history of the past anew with reference to the conditions uppermost in its own time," Frederick Jackson Turner famously wrote in 1891. The goal of historical inquiry, he asserted, was not to evoke "the dead past" for its own sake, but to use the past as a means to illuminate "the living present." His great concern was the origin and development of American democracy, which, by the 1890s, seemed to have reached a turning point, if not a crisis. In his judgment, as he explained in his even more famous frontier thesis essay of 1893, the vast landscape open to European American settlement—the environmental incubator of a distinctive national democracy—was no more. "And now," he concluded, "four centuries from the discovery of America, at the end of a hundred years of life under the Constitution, the frontier has gone, and with its going has closed the first period of American history."

The early twenty-first century can be imagined as a Turnerian moment in the study of American nature preservation. Global climate change is prompting historians to write about the past anew with reference to a living present that is becoming acutely aware of its perilous future. A period of American development seems to have gone, and another chapter seems to be opening. On the cusp of the Anthropocene, the past looks different. With a view toward an unsettling future, an overlooked story of nature preservation—a story with democracy at its heart—might be worth examining.

Contemporary Americans often think of nature preservation as if John Muir and the Sierra Club invented it, as if it applied only to monumental landscapes, wilderness, wild things, and ecological and evolutionary processes. Its roots actually run deeper. Muir and the Sierra Club indeed were American originals, but their originality derived less from their creation of a new concept than from their reinvention of an older one. Earlier Americans thought and talked about preservation as it pertained to human nature, human concerns, and human political prospects, especially democracy. From the American Revolution through much of the nineteenth century (a period that roughly corresponded to Turner's "hundred years of life under the Constitution"), people often spoke of preserving human nature and the natural rights that guaranteed individual self-determination. From the nineteenth century into the twentieth, some Americans called for preservation in national parks and other protected areas not simply for nature's sake, but also as a means to strengthen the democracy that the Revolution made possible.

Even as preservation retained a vital tie to the Revolution and other upheavals that defined the nation, a remarkable cultural break occurred. While Turner looked back on the vanished frontier, some of his fellow citizens severed preservation from its human purposes. The older worldview lived on, and historians have made a compelling case that some people extended the concept of natural rights to forms of nonhuman nature. For the most part, however, the new version of preservation lost its application to humanity, and with it, its democratic entailments. From that moment onward, the idea that preservation implied a kind of purity apart from humanity began to spread. If preservation is to have any utility in the twenty-first century, if it is to be something more than an outmoded romantic ideal unsuited to a human-dominated world, then citizens might want to recall its origins and its service to the nation's democratic promise. That side of preservation, at least, might be worth carrying into the brave new world of the Anthropocene.

An eighteenth-century notion of preservation informed the American Revolution. As John Locke, a philosopher favored by many revolutionaries, wrote, "whenever any one shall go about to bring them into such a Slavish Condition, they will always have a right to preserve what they have not a Power to part with; and to rid themselves of those who invade this Fundamental, Sacred, and unalterable Law of *Self-Preservation*, for which they entered into society." More simply, as proponents of natural rights often said, self-preservation is the first law of nature. Although the Declaration of Independence did not contain the word "preservation," it invoked "the laws of nature" and outlined the colonists' need to preserve themselves and their liberties from the depredations of an unnatural, tyrannical monarchy.

The victory in the Revolutionary War broadened preservation's emphasis from the self to the nation at large. Jefferson, Madison, Franklin, and others believed that western lands and waters were essential to the preservation of the republic. The West would keep the United States in a youthful, vigorous condition, by absorbing population, democratizing wealth and defusing class conflict, and providing the means to end slavery. Although the Constitution did not mention preservation or nature, it was a future-oriented, process-centered document that allowed the nation to absorb new land and create new states, and enabled its white male citizens to transform land and resources into property and wealth, the basis of economic independence and the guarantor—the preserver—of their political virtue. The settlement of the West, to which Jefferson looked forward, was the same process upon which Turner looked back and interpreted as the generator of a frontier democracy.

Before, during, and after the Civil War, the idea of preservation flourished, perhaps even experienced a golden age. The manifestations bloomed like wildflowers after a fire that liberated nutrients and a rain that nourished sprouts. Americans called for the preservation of liberty, the Union, the Confederacy, access to western lands, fisheries, and forests. Patriots called for the preservation of battlefields, cemeteries, and the homes of founders and heroes. The Mount Vernon Ladies Association believed that the preservation of a common symbolic past—the home of George Washington—would help rescue the Union. Black Americans struggled to preserve their physical selves and, eventually, the official recognition of their rights. President Abraham Lincoln issued the Emancipation Proclamation of 1863, which relieved a sizeable portion of human beings from the commercialization of their bodies and helped keep the United States intact. Although this was a fraught era, its many preservation movements revealed a vibrant link between nature and democracy.

The first national parks appeared in the midst of this fecund preservationist landscape. In 1832, the Englishman George Catlin called for the preservation of wild nature in the form of a park. Henry David Thoreau and others made similar appeals. Lincoln honored this impulse in 1864 when he signed into law the Yosemite Park Act, which reserved a portion of land in the Sierra Nevada from private ownership. In 1872, Congress and President Ulysses Grant set aside Yellowstone National Park and required the secretary of the interior to issue "regulations . . . for the preservation . . . of all timber, mineral deposits, natural curiosities, or wonders within said park, and their retention in their natural condition." Both parks' legislation established their democratic purpose. California must hold the Yosemite grant "for public use, resort, and recreation," and Yellowstone existed "for the benefit and enjoyment of the people."

Although preservation precluded the privatization of protected lands by small freeholders, ideologists stepped forward to reconcile the parks with a democratic purpose. Among them was Frederick Law Olmsted, an opponent of slavery, a founder of landscape architecture, a designer of New York City's Central Park, and the author of a statement, penned in 1865, on the meaning of Yosemite and the Mariposa Grove of Sequoia trees. In that document, Olmsted aligned the parks with the values of a nation that had just endured its greatest crisis. The role of government, he said, was to protect "all its citizens in the pursuit of happiness." In the same spirit, the "duty of preservation" incumbent on state and nation had rescued Yosemite from privatization so that the "health and vigor" of the entire citizenry would benefit from the "free use" of a wondrous mountain landscape. Olmsted's concept was not quite the same as the idea that the West would keep the republic from decay, but the reinvigorating function of a beautiful western park would serve much the same end.

The association of preservation and democracy continued into the twentieth century. Elements of it persisted in the work of activists such as Robert Marshall, a socialist committed to the democratic experience of wilderness, in the writings of Aldo Leopold, in ideas and slogans such as "parks for the people," in the efforts of grassroots activists to rescue the landscapes that they loved and that contributed to their material well-being and informed their identities as citizens. Anywhere and anytime Americans argued for a link between preservation and democracy, they echoed ideas that Olmsted had articulated.

To point out that nature preservation had a democratic function is not to claim that it was free of problems. Early parks and other protected areas often excluded rural and working class people, immigrants, and Natives from resources that they needed and to which many had rightful claim. Preservation often served as a means for the wealthy and the powerful to marginalize small producers and turn national parks and other areas into special places for the privileged few. If democracy is the notion of multiple voices shaping public policies and spaces, then the first acts of nature preservation also expressed exclusive if not elitist purposes.

As the nineteenth century gave way to the twentieth, some Americans began to think of preservation without any connection to democracy at all, as the protection of nonhuman nature for its own sake. Signs of this shift appeared in the life and writings of John Muir. Yosemite certainly gave Muir pleasure and made him happy. Perhaps his stirring, eloquent essays exemplified Olmsted's hope that the experience of the park would elevate the health and vigor of the people's intellect. Nonetheless, in Muir's writings, preservation lost a vital tie to democracy. He acknowledged the hunger of "the people" for meaningful contact with nature, but mostly he was indifferent if not opposed to the idea that preservation should serve a human purpose. Nonhuman nature and its protection was his concern.

Indeed, in "the common dust-and-ashes history of the public domain," as Muir put it, in a city's quest for a water supply, and in the actions of sheepherders, lumberjacks, hunters, miners, and countless others, democracy was the problem. If anything, Muir asserted, the people had an obligation, without regard for self-interest, to protect nonhuman nature simply because of its intrinsic value. "These kings of the forest, these noblest of a noble race," he wrote of his beloved Sequoia trees in a telling use of metaphor, "rightly belong to the world, but as they are in California we cannot escape responsibility as their guardians. Fortunately the American people are equal to this trust, or any other that may arise, as soon as they see and understand it."

Muir certainly had reason to think as he did. The Revolutionary legacy simply was not as important to him as it was to other Americans. Although opposed to slavery, he neither campaigned against it nor served in the military, so the Civil War was not, as it was for Olmsted, Lincoln, Frederick Douglass, John Wesley Powell, and many others, a defining struggle. And to be sure, the Revolution's high-minded ideals often dissolved in the nineteenth century rush to exploit the public domain. What was Muir to think of overgrazed meadows, shattered forests, dusty pastures, mountainsides scarred by mining, ghostly prairies devoid of herds, and canyons that engineers could imagine only as reservoirs? Exclusion from national parks did not necessarily make heroes of small rural producers, and how was Muir to judge the workingmen who, in the spirit of the gold rushers of an earlier day, announced their plans to move from one easily exploited resource frontier to another, from shingle splitting to duck hunting to grazing? Muir's landscapes were a long way from the exalted ideals of Thomas Jefferson.

The upshot is clear. Much as Turner defined a watershed in the evolution of American democracy, so did Muir, at virtually the same time. To preserve nature no longer was to serve the Revolution's end, but to defend God's sacred temple from the common lot of greedy human despoilers. Muir alone was not responsible for this separation of human and nonhuman nature. A primary feature of modernist thought was a reductive separation of things — of races, genders, ages, cultures, nations, and species from one another, and ultimately of people from the natural world. Whether in minds or on the ground — in animal cages and zoos, in systems of Jim Crow segregation, in the roundup of Native tribes and their confinement on reservations, in the hardening of national borders, or, indeed, in the creation of national parks — separation gained force. Academic historians furthered the trend, and they began to tell stories about the human struggle for self-determination as if that struggle bore little or no relation to the inherent material qualities of the earth — as if preservation only mattered to Sequoias or bison, as if democracy and nature never had been in dialogue.

Turner himself imbibed of the older worldview while fashioning the new. He never used the word "nature" to describe the context in which American democracy developed. History, he wrote, "is past literature"; "it is past politics, it is past religion, it is past economics. . . . History is the biography of society in all its departments," although evidently no department, in his conception, had a door that opened onto nature. He acknowledged that American democracy arose from European Americans' interaction with a wilderness environment and its "free land," but beyond that, nature had no place in his narrative. Looking back on a history in which land had been integral to the development of the United States, Turner wrote from the perspective of a present that increasingly imagined the struggle for democracy as something above and beyond the natural world.

This tension—in some cases in outright conflict—between preserving nature for democratic ends and preserving nature for its own sake continued through the twentieth century and into the twenty-first. It was evident, for example, in the 1964 Wilderness Act, which preserved wild areas "for the use and enjoyment of the American people" but "where man himself is a visitor who does not remain." It was evident in the deep ecology movement of the 1990s, which asserted the primacy of the ecological community—of which humans were part and from which they should benefit—over the anthropocentrism of Western civilization. It was evident in the desperation of people not inclined to turn their backs on their fellow citizens, but who wanted to honor a moral imperative to rescue animal and plant species threatened with extinction.

If preservation is to have any use in the Anthropocene, it must revive and deepen its democratic roots. The act of saving and protecting nature need not be imagined as serving only a separate, nonhuman end. To the contrary, preservation offers an opening for citizens to work together in ways that can fulfill the democratic promise fostered in the Revolution, restated in the writings of Olmsted, and celebrated in Turner's history. The noble act of defending and perpetuating nonhuman nature can help revive and ennoble human community and its democratic practices. The prospect of the Anthropocene is dreadful, but the crisis also presents an opportunity to shake loose ossified historical memories and exercise atrophied political skills. As in the past, people's feelings for all that they hold dear—for the community of life, for plants and animals and places, for democracy itself—are never so strong as when they are under threat. Like other republics, this republic has risen to challenges in the past, and it has the potential to do so again.

To protect cherished things is among the deepest of human urges, and a word and concept of such extraordinary emotional

Figure 15. The preservation of nature presents an opportunity for citizens to cultivate skills in communication and deliberation essential to the preservation of democracy. Parks as Portals to Learning, interdisciplinary field school at Rocky Mountain National Park, August 2013. Photo credit: Maren Bzdek, Public Lands History Center, Colorado State University (used with permission).

power is almost certain to remain in use. But if Americans wish to rescue those things that matter most, if they wish to withstand the heat and the storms that are sure to visit them, if, in other words, they wish to preserve themselves, then they might ponder and deliberate over what remains of their democracy as one era passes and another begins. In a plurality of voices united in an ongoing process of experimentation, deliberation, and revision, twenty-first-century preservation must transform the past into a living present in which a cherished natural heritage is at the heart of any decent conception of America.

Green Fire Meets Red Fire

Stephen J. Pyne

William Shakespeare famously claimed the past was prologue. Henry Ford replied it was bunk. To the more ardent Anthropocenarians the present has so ruptured from previous times that the past can offer no meaningful guidance. This sentiment can free-associate with notions from natural science and philosophy, both of which claim to stand outside human culture and history. Taken together they suggest that we face a no-analogue future for which our only recourse is to reason from first principles and transcendent concepts and to respond on a scale commensurate with the threats. We have to think big, and we can't rely on the past to inform our understanding.

I disagree. While ecology might have some general laws, it is a historical science, and while nature preservation might bid to rise above its social setting, it is a historical moment, an idea whose time and place can come and maybe go. The Anthropocene is a construct that defines itself against what came before. The past has a way of reasserting itself. So even when the past has been charbroiled to ash or slow-cooked to oblivion in greenhouse gases it returns because, in truth, it never truly goes away. Or to bookend Shakespeare with William Faulkner, the past isn't over, it's not even past. There is good cause to view the uneasy détente between nature preservation and the Anthropocene historically.

As a useful index, consider that we are the keystone species for Earth's fire. We hold a species monopoly; it's what we do that no other creature can. Our stewardship of flame makes a unique metric of our ecological agency and of how we have sought to preserve nature. Or to place that observation more concretely, consider one of the hallowed sites of American environmentalism, the high rimrock on the Apache National Forest where Aldo Leopold shot a she-wolf, watched the "fierce green fire" in her eyes die, and then pondered what it might mean to think like a mountain. In 2011 a megafire boiled out of the Bear Wallow Wilderness and overran that scene, along with 538,000 acres generally, to become the largest forest fire in Arizona history. In that fierce red fire American preservationist thinking met the Anthropocene.

*   *   *

For a fire historian the idea of the Anthropocene is both easy and awkward to accept.

The easy part is that fire has been used to define the era, which begins when humanity shifted from burning surface biomass to burning fossil fuels. That "shifted" is the awkward part because hominins have been tinkering with fire since *Homo erectus*. Our pact with fire is our original Faustian bargain. We've always burned and so have affected the landscapes around us, and since we have gone everywhere, that pretty much means the landed fraction of Earth (save Antarctica). And in adopting fire we cooked food, and precooked landscapes, in ways that have altered our genome.

Why might this time be different? One reason is scale — the range and burn rate of human-brokered combustion. With industrial fire we've extended our reach, if not our grasp, to the atmosphere and oceans, and across deep time. We're burning landscapes from the geological past and releasing their effluents to the geologic future. Even Antarctic ice is affected. Domestica-

Figure 16. The Leopold wolf-kill site, looking south, amid the wind-driven Wallow Fire. The mixed-conifer mountain in the distance was severely burned, the ponderosa pine of the rimrock burned patchily, and the gorge selectively. Burn severity, in brief, matched fuel loading and exposure to free winds, with the highest and most densely forested lands burned harshly and the secluded ravines lightly touched by fire. Photo credit: Steve Pyne.

tion by fire, and by other less homely means, has so evolved that not only are species being lost but select genomes may be engineered into new expressions, other genomes may be invented, and there are yearnings, by some, to resurrect formerly extinct species. A destabilization of the planet's genomes may match that of its geographies. The magnitude of this biotic alchemy across the planet may signify a qualitative phase change.

It is not just that a hypothetical "balance of nature" has vanished, but the sense of a natural order that can exist outside of humanity. Our burn rate is stirring natural and anthropogenic combustion into a common cauldron. There may be no autonomous referent against which to measure ourselves; the primary agent for destabilizing Earth is destabilizing itself as well. But the wobbling gets worse. It may apply to the story we tell as well, or our ability to understand and describe these changes. We are both actor and critic. We have met the Other, and it is Us. Our story may turn on itself like a Möbius strip.

*   *   *

For a historian of the American environment the narrative of nature protection hinges on the presence of a public domain.

Their existence derives from two paradoxes. One is that industrial societies—those most ravenous of natural resources—are also the ones prone to create nature preserves on a significant scale. The other is that a relatively unbridled capitalist society, in the full flush of the Gilded Age, set aside roughly a third of its national estate to shield against the ravages of its own economy. Barring a global pandemic neither event is likely to repeat.

In effect, finding it too difficult to extract public goods out of private lands, the United States shifted its environmental commitment to the public domain. Like a shaky banking system that gathers up its debts into a bad bank to free up the good banks to function, so the country created good lands to absorb the duties it was unable to do on bad ones. There are some powerful stories of public action against toxins like DDT, contaminated rivers and shorelines, and befouled air, but the narrative of American environmental exceptionalism hinges on its preserved wildlands. It's the public lands that require environmental impact statements, that absorb the duties of the Endangered Species Act, that house legal wilderness, and that hold the great epics. Pollutants are a task of national housekeeping, but the wild is romantic, and by being uninhabited in any serious way, it is where grand schemes for restoration, rewilding, or other projects advancing non-anthropocentric goals can be imagined.

Their history begins when they were isolated from the national estate. A couple of wondrous national parks were first, but the bulk of the lands were national forests. Wildlife refuges and national monuments added variety. Not until the Taylor Grazing Act of 1934 was the public domain closed to further private acquisition; at the same time there was some transfer from private to public holdings through purchase and abandonment. The heroic age was about halting the divestment of the public domain and establishing agencies to oversee the reserves. Similarly, the prevailing fire policy was to exclude flame as far as possible and so spare the reserved lands from the abusive burning common outside them.

By the 1960s this era was closing. The coming controversies were over expanding that land base by purchase through the Land and Water Conservation Fund and, especially, over internally segregating the public domain for different purposes. Here was the basis for the great political battles over dams in the national parks, the application of the Wilderness Act (1964), and the use of the Endangered Species Act (1973) to create preserves indirectly. Meanwhile, The Nature Conservancy (TNC) began its major acquisitions, putting technically private lands to public purpose and in many cases acting as a broker for an ultimate transfer to public status.

This sorting out has been an informing narrative for contemporary environmentalism. Every federal agency either received a charter for the first time (like the Bureau of Land Management) or had its charter renewed; even the Forest Service went through a fast march of legislative mandates. The defining quarrels occurred when the private economy tried to penetrate the reserved domain in the form of logging, mining, grazing, and mass-development recreation, and more broadly those quarrels sparked as public lands were redefined from generic multiple use to specific purposes. Outside the public lands there was little effort to accommodate nature protection apart from a handful of states.

Fire policy obligingly shifted from a singular strategy to one that adjusted practice to each category of land use. Agencies sought to restore natural fire to wilderness, shield developed areas from any flame, and reinstate tamed fire elsewhere through prescribed burning. A taxonomy of free-burning fires emerged according to the category of lands burned and the source of ignition, natural or human. Fires were wild, prescribed, management, planned or unplanned, prescribed natural, wildland fire use, or resource benefit — a managerial menagerie with each label carrying its own obligatory protocol. Interagency consortia replaced a once-hegemonic Forest Service.

Now that era seems to be expiring. The population of the country doubled between 1950 and 2000; sprawl has carpet bombed the countryside with houses, all interbreeding with whatever natural hazards are present; ecological diversity is coming to mean the varieties of suburbs and exurbs and the fragmentation they have wrought. Sprawl, not commodity production, threatens the integrity of the reserves' borders. But the primary controversies hinge on the internal administration of the public domain, particularly the trend toward ecosystem management.

The public lands appeared overrun by an ecological insurgency. They are, by most accounts, a shambles from poor practices in the past, unhealthy biotas, invasive species, beetle and budworm swarms stripping conifer forests on the scale of the Laurentide ice sheet, fast-morphing climate, and fires. The founding model — set the lands aside from the ravages of footloose capital and folk migrations — is no longer sufficient. Nor is its replacement: segregate the lands by their relative pristineness. It does little good to set lands apart for special protection if they rot away from the inside. Wildness will survive, wilderness might not.

Now the country is inflecting into another phase. The future will continue to depend on the public domain, much as the country's infrastructure relies on national networks and federal funding. But the goals of nature protection must lie beyond expanding the federal lands and arranging them by degrees of purity.

The future of the protected land base lies with making sprawl more eco-friendly, with efforts by the states to create protected sites, and with private landowners, among which TNC can claim a niche both unique and large.

Classic preservation is segueing into restoration. Instead of sprawl projecting itself into the wild, smarter landscaping could carry a sense of a natural world into human communities. The future of management on those sites will rely on using national resources as a fulcrum to fashion collaborations and conservation easements, and to leverage lots of small changes into landscape-scale or regional projects. The future will probably look like the Disney Wilderness Preserve managed by TNC, the wholesale Everglades restoration program, the Four Forests Restoration Initiative in Arizona, and the New Jersey Pinelands Commission.

All depend on active administration. That doesn't mean phalanxes of chain saws and feller bunchers, or the ecological equivalent of broadcast antibiotics that kill the good along with the bad. It means more targeted, better informed interventions. The alternative is that wilderness and parklands may simply be overrun by the Pandora's box of global change opened by the Anthropocene. The old preservationist ideal that nature preserves should remain as sacred groves, untouched so far as possible, is the formula of a faith-based ecology. It can satisfy spiritual yearnings. But symbols need not be big, and preserves may no longer be able to serve as ecological anchor points. Paradoxically, that task may fall to their nominal buffer zones.

A pragmatic protectionism suggests that some ecological engineering is called for to ensure biological goods and services. In the nineteenth century the state had to halt the wreckage by placing some lands outside the reach of global capital; in the twenty-first it will have to intervene within those reserved lands to prevent the reach of the Anthropocene. This is a more complicated task. The choices before us are not aligned along a spectrum between the two poles of Prometheanism and primitivism but arrayed like a constellation imagined out of the infinite lights of the night sky.

So, once again, fire policy is morphing. The carefully parsed categories have become meaningless when megafires can gallop across landscapes a hundred times the size of the minimum wilderness and burst beyond the borders of the reserves into Colorado Springs and Bastrop County. The effort to substitute prescribed fire for wild fire has succeeded in the Southeast, where a cultural tradition of controlled burning on working landscapes has endured, but it has faltered in the West, where the working landscape has shriveled. Instead, agencies are taking an "appropriate strategic response" to every fire. They are scrapping the distinction between natural and human causation. They treat wildland fires as big-box events in which crews back off to defensible barriers and burn out, while offering point protection to assets like houses.

The emerging model is a hybrid of government and civil society, much as agencies hybridize natural and anthropogenic fire. The interagency theme has expanded into an inter-governmental one in which all jurisdictions, both public and private, must coordinate under a "national cohesive strategy." NGOs like the Coalition of Prescribed Fire Councils have arisen to promote fire's restoration on private lands. TNC now prescribe-burns as much each year as the National Park Service. The critical environs will likely involve both public and private lands; their guiding principle, a new version of the working landscape, one dedicated to ecological benefits. The outcome depends on not just what is done but how and at what scale. As Paracelsus reminded us during the Renaissance, toxicity resides in the dosage, not the substance.

What has changed are not the physical principles of fire behavior or transcendent concepts of naturalness but historical circumstances. The need to reconcile notions into practice means people must make choices in a contingent world about which they have incomplete knowledge. Ideas can no more flourish in untrammeled preserves than can fires. This sounds a lot like pragmatism.

*   *   *

For a historian of human affairs the problem is not that the past offers no analogues but that it holds too many.

The hope that history might serve as a reference baseline or a repository of lessons and techniques is naive. What the past offers is less technical assistance than moral guidance. The core of the Anthropocene remains the human hand guided by mind and heart, and its moral challenge remains, as William Faulkner once observed, "the human heart in conflict with itself."

The fact is, technology can enable but not inform, and science can inform but not choose. Some choices we know have little redeeming social or ecological value, and for a few we already know enough to act on. Some past choices made with the best science of the day have proved wrong, and while science can self-correct over time, the damages done may not. So it is with ideas that suit one era but are maladapted to another. Stick a torch into forest duff today and you are not likely to re-create a presettlement burn but a blowup.

A reading of history can serve the future by sketching the historic ranges of moral variability, as it were. It can illuminate how to cope with uncertainty and ambiguity; by demonstrating how to act—which the future requires—with some degree of prudence and urgency; and by shunning the false dichotomy that either would leave land, air, and sea untouched in the hopes that a Panglossian best of all possible worlds will self-emerge or would demand interventions on the same scale as those by which we've unhinged the planet. History's enduring lessons testify less to techniques than to such virtues as humility, resilience, endurance, courage, tolerance, and the value of pluralism.

Or to return to our fire analogue, we will be caught between two fires, one of industrial combustion that underwrites the Anthropocene and one of free-burning flame that epitomizes untrammeled nature. How they express themselves are not simply telling analogues but trying flames. We must somehow pass between green fire and red without turning everything black.

Restoration, Preservation, and Conservation

AN EXAMPLE FOR DRY FORESTS OF THE WEST

William Wallace Covington and Diane J. Vosick

The frequent fire forests of the West present a unique opportunity to return resilience to forest ecosystems whether they are in the wildland-urban interface or a wilderness area. However, two misconceptions often challenge restoration efforts. The first is the attitude that ecological restoration that relies on the removal of trees is contrary to preservationist goals — that attitude continues to impede restoring public lands at the pace and scale that is needed to counteract the threat of unnatural fire. The second is that restoration that is based on the historic range of variability is irrelevant in the face of climate change. We argue that not only is restoration consistent with preservation goals that seek to sustain intact ecosystems but that it is also improves forest resilience and health that will position these systems for the uncertainties of climate change.

Restoration Principles

Ecological restoration offers a practical approach for developing scientifically and ethically sound conservation management — management that not only treats symptoms of ecosystem health decline, but also attacks the underlying causes of such decline. Restoration is essential for degraded areas set aside for preservation of nature (e.g., preserves, natural areas, or wilderness), but it is also a desirable goal in the management of lands where nature conservation and resource use are shared goals. Finally, scientifically rigorous and collaboratively designed restoration plans can result in common ground that can potentially reconcile preservation versus use conflicts.

Ecosystem restoration is founded upon fundamental ecological and conservation principles and involves management actions designed to facilitate the recovery or reestablishment of native ecosystems. Over evolutionary time, species not only adapt to their evolutionary environment, but they may also come to depend upon those conditions for their continued survival. Thus, the greatest threat to biological diversity is the loss of evolutionary habitats, and the greatest hope for reversing the losses is restoration of these habitats and their self-regulating processes. This leads us to a central premise of ecological restoration: that restoration of natural systems to conditions consistent with their recent evolutionary environments will prevent their further degradation while simultaneously conserving their native plants and animals, especially in the face of climate change.

Although contemporary climate changes are accelerating the need for adaptation, a healthy natural system will have much of the intrinsic natural resilience and genetic resources required to respond adaptively to change. A useful analogy is human health. Formerly ill people whose health has been restored (to within the natural range of weight, blood pressure, etc.) are much more likely to be able to withstand stresses than are those whose health has not been restored. Restoration positions forest ecosystems to confront uncertainty in the best condition possible.

Practitioners of ecological restoration recognize that a failure to include human interactions with restored systems is not only unrealistic, but also undesirable for their long-term sustainability. In fact, in cases where novel conditions prevent natural system functions, ongoing management may be required to compensate for the unnatural conditions. Examples of such a circumstance are where restored sites are too small to support

Figures 17 and 18. Gus Pearson Natural Area, Fort Valley Experimental Forest, Coconino National Forest, Arizona, before and after restoration (1992 and 2004). These two photos demonstrate how restoration (a combination of thinning and burning) can take forests that are severely departed from natural conditions and set them on a trajectory of recovery. This treatment was designed using reference conditions. Photo credits: Wallace Covington.

natural predator-prey dynamics or to accommodate natural disturbance regimes such as natural fire.

Ecosystem restoration should not be construed as a fixed set of procedures, nor as a simple recipe for land management. Rather, it is a broad intellectual and scientific framework for developing mutually beneficial human and wildland interactions compatible with the evolutionary history of native ecological systems. In other words, ecosystem restoration consists not only of restoring ecosystems, but also of developing human uses of wildlands that are in harmony with the natural history of these complex ecological systems.

Ecological Restoration: A General Scientific Framework

Ecological restoration is the restoration of natural ecosystem structures and processes, including self-regulatory mechanisms. Treatments are based on reference conditions (the evolutionary environment context) and consistent with evolutionary and conservation biology and ecosystem ecology principles.

SOCIAL AND POLITICAL FRAMEWORK

In an ecosystem ecology approach, social and political concerns play a major part in defining treatments. So stakeholders must be engaged, especially community-based partnerships linked to regional and national agencies and interest groups, with policy makers, natural resource specialists, and resource managers.

OPERATIONAL FRAMEWORK

Financial and personnel constraints place geographical limits on treatments. Emphasis is placed on strategically located restoration fuel breaks that are anchor points for large, landscape-scale treatments. These breaks can protect key landscape ecosystem components such as human communities, critical habitat for threatened or endangered species, and core areas of greater ecosystems such as wilderness areas and national parks.

ECOSYSTEM-MANAGEMENT FRAMEWORK

Restoration and fuel-reduction goals should be integrated with overall ecosystem conservation and management goals. Reference conditions serve as a starting point for the goal of scientifically based land-management objectives.

ECONOMIC FRAMEWORK

Economic analyses should consider all costs and savings. Restoration-based fuel treatments save money by avoiding the costs of firefighting, rehabilitation, and compensation for property damage. They are also an investment in protecting lives. They present new opportunities for rural economic development through restoration-related jobs and products. Ecological economic analysis will probably indicate that benefits greatly outweigh costs.

ETHICAL FRAMEWORK

We have a responsibility to future generations to solve ecosystem health problems. Ecological restoration speaks to the land ethic—humans should be good stewards and show a caring concern for nature.

Ecological Restoration: Getting to Landscape Scale

Concern about the degradation of public lands and associated natural resources began in the 1970s. Disruption of natural disturbance regimes and the overutilization of resources are especially disruptive in dry forests, particularly those once dominated by open ponderosa pine forests. Over the past two decades we have witnessed sudden leaps in aberrant ecosystem behavior long predicted by ecologists and conservation professionals such as Aldo Leopold and Harold Weaver. Trends over the past half-century show that the frequency, intensity, and size of wildfires have increased by orders of magnitude—and along with it are the loss of biological diversity, property, intact watersheds, and human lives. Emerging research shows that in some places where fire has burned severely and where conditions are hot and dry the ponderosa pine forest is being converted to a shrub-woodland—blowing holes in what once was the largest contiguous ponderosa pine forest in the world.

The new reality of severe, landscape-scale (mega) fire and postfire flooding, when considered with the predicted effects of climate change, has led to new efforts to tackle restoration at the pace and scale of the problem. Policy makers, land managers, and stakeholders are working together to develop new strategies to accelerate ecological restoration. Recent efforts include passage of the Collaborative Forest Landscape Restoration Program (CFLRP) designed to encourage collaborative planning and restoration at the landscape scale. To date there are 23 landscapes included in the program. Taken together they are forging new ways to work at scale.

Working at the landscape scale will also require a fresh look at the laws, policies, rules, and regulations that guide federal land management. Many policies derive from periods where societal goals for public land management were different than they are today. Some of these policies are no longer relevant or are written in a way that didn't foresee current environmental conditions such as climate change and megafire. As our understanding of natural systems has expanded and societal desires for public lands have evolved, old and inappropriate legal frameworks undermine restoration and work at cross-purposes for effective land management.

From our perspective, whether it is the collaborative design of treatments or new policy directives, land managers should build from solid science and the aforementioned ecological principles; they should embrace a long-term ecological perspective and *then* consider social and ideological issues. When management tries to balance ideological concerns, the ecological trade-offs should be carefully considered. That's not to say it's inappropriate to manage for multiple objectives, but rather that the implications for achieving resiliency and long-term ecological health should be fully acknowledged.

A close look at the Four Forest Restoration Initiative (4FRI) illustrates how an innovative approach to achieving landscape-scale restoration embraces restoration principles while navigating between old and new land-management objectives and evolving societal desires for public lands. Although the 4FRI has a unique history, the challenges and opportunities unfolding in this project are shared by other landscape-scale projects throughout the West.

A Case Study:
The Four Forests Restoration Initiative

One of the largest and most ambitious applications of ecological restoration in western forests is the 4FRI in northern Arizona. The area encompasses 2.4 million acres across four national forests that are dominated by southwestern ponderosa pine. The 4FRI began in 2009 as a step toward achieving the landscape-scale restoration goals envisioned in the "Statewide Strategy for Restoring Arizona's Forests" developed by then Governor Janet Napolitano's Forest Health Advisory and Oversight Councils. The 4FRI immediately attracted national attention and received a funding boost of between $2 and $4 million per year when it was chosen as a pilot project under the 2009 Collaborative Forest Landscape Restoration Act (CFLRA). This project did not spontaneously coalesce in 2009 but rather stands on the shoulders of 100 years of scientific and collaborative effort.

In 1908 the first forest research station in the nation was established on the Coconino National Forest. Scientific research in the area spans more than 100 years, generating a substantial body of peer reviewed research that contributes significantly to our current understanding of southwestern forest ecosystems.

As far back as the 1940s ecologists observed that forest structure was changing as a result of fire suppression. As trees irrupted and grew in historically unprecedented numbers, concerns arose about the potential for severe fire and unexpected ecological changes. Experiments designed to understand the role of fire and the effects of restoration treatments were begun in earnest in the 1970s. These studies and the subsequent and ongoing work of forest ecologists and wildlife biologists provide the *scientific framework* for restoration for southwestern ponderosa pine forests.

During the 1990s forest management in the Southwest became a lightning rod for environmental activism. Concerns over declining populations of the Mexican spotted owl (MSO) and the impact of commercial logging on northern goshawks led to the listing of the MSO as federally threatened, the shutdown of 11 national forests, and widespread challenges to forest management. By the late 1990s mechanical harvesting of trees was at a standstill. Concurrently, forest conditions became ripe for unprecedented crown fire. The trees that gradually filled in the forest over a span of 80 years because of fire suppression were sufficiently tall and dense to carry fire into the crowns of trees and move over distances at rates of speed never witnessed in contemporary times.

Fear of fires and frustration over lawsuits led to the formation of collaborative groups comprised of diverse interests with the goal to identify socially acceptable restoration treatments. In the Southwest this *social and political framework* had mixed success deflecting legal challenges.

In the 1990s ecological restoration was a new management paradigm. Replicated studies and demonstration plots using mechanical removal of trees to achieve restoration were viewed by some with skepticism. Distrust of the science and the Forest Service undermined efforts to find common ground. The Forest Service during that time and even today still struggles to identify its own proper role in the collaborative process and the boundaries to its ability to adopt socially developed management agreements.

The 4FRI Stakeholder Group struggles to balance the demands of litigious environmental organizations within the constraints on collaboration and treatment design imposed by the Forest Service. Meanwhile, litigious environmentalists who have never visited the 4FRI nor worked with the collaborative maintain their standing to challenge the final 4FRI environmental impact statement (EIS) required under the National Environmental Policy Act (NEPA) if they have commented earlier in the environmental review process. Although restoration science doesn't support imposing a diameter cap to limit the size of trees to be removed during restoration, litigious environmental groups have argued for diameter caps and similar assurances in order to prevent the Forest Service from removing large trees. Recent federal forest policy embraces collaboration as the way to accelerate treatments and lower the probability of litigation. Whether or not collaboration will lead to accelerated restoration across the 4FRI landscape hinges on whether or not the first one-million-acre EIS analysis is successful.

Applying an *operational framework* consistent with ecological restoration principles is a challenge for the 4FRI on multiple fronts. Long planning horizons related to NEPA and congressional policy directions constrain the federal land-management agencies' ability to direct treatments to the areas of most critical ecological need or to where treatments may play a strategic role in modifying fire behavior. For example, the first collaboratively designed EIS for the 4FRI was scheduled to be completed in summer of 2014, four years following the designation of 4FRI as a pilot project under the CFLRA. In the meantime, projects are using previously approved projects — those that already have NEPA records of decision record of decision. Some of these projects only minimally reflect current approaches to ecological restoration. The Healthy Forest Restoration Act of 2003, as well as follow-up language in appropriations bills and directives from the Office of Management and Budget instruct the federal land-management agencies to put treatments in the wildland-urban interface (WUI). Although smart for protecting human assets at risk, these treatments fail when it comes to addressing the underlying problems of degraded forest ecosystems and associated severe, landscape-scale (mega) fire. Placement of treatments to protect critical habitat manifests an ironic twist when it comes to preservation versus restoration. The latest recovery plan for the MSO identifies unnatural, severe fire as the greatest threat to MSO habitat. Yet the environmental community has not argued for treatment placement designed to protect critical habitat. In fact, just the opposite has occurred — they have supported focusing treatments in the WUI. Out of fear of litigation by environmentalists, the Forest Service has shied away from focusing treatments to protect one of the most irreplaceable ecological assets on the landscape — critical habitat such as old growth and MSO habitat.

One hundred years of land-management policies from the original 1905 charter for the Forest Service to policies that encourage grazing, mining, endangered species protection, and sustained timber yield, as well as a mosaic of other, often competing objectives, make it extremely difficult to implement an *ecosystem management framework* in the 4FRI landscape. In addition, many of the national forest plans developed under the 1976 National Forest Management Act were written in the 1980s and don't even acknowledge the existence of ecological restoration.

Experience shows that it is possible to design and implement treatments using reference conditions and the natural range of variability; however, managers have been reluctant to do so. Marking crews need additional training that is different from conventional approaches. The Goshawk guidelines that are the default framework for most acres across the 4FRI forest call for a regulated approach to forest structure that undermines restoration of natural conditions. The demands of recreationalists including off-the-road vehicle users, bicyclists, and hikers require the Forest Service to maintain roads and trails that can disrupt wildlife and impair watershed function. Providing the multiple uses that citizens demand of their forests makes managing for ecologically resilient intact landscapes difficult.

A central goal of the 4FRI collaborative is the *economic framework*. Historically, forest industry was a major employer and economic driver in rural northern Arizona. In the late 1980s and early 1990s it faltered and then collapsed. Economic analyses demonstrate that at full operation the 4FRI should generate over 440 full-time equivalent jobs in the harvest, hauling, and manufacturing sectors — jobs that have the potential to provide middle-class wages in an area of the state with some of the highest unemployment in Arizona. Equally significant are the economic values attached to a healthy forest. Tourism is important to the state, especially to the small rural communities embedded in and around the national forests. The value of the water resources, that originate in the mountains within the 4FRI landscape and contribute a significant amount of water to Phoenix and surrounding valley cities, has attracted new attention from Phoenix and nearby municipalities and the Salt River Project. These stakeholders are now exploring innovative approaches to increase the financial and human resources available to accelerate restoration and avoid catastrophic fire.

Finally, a recent study of the full cost of the 2010 Schultz Fire and subsequent flooding demonstrates the value of investing in the prevention of severe, unnatural fire. The 15,000-acre fire near Flagstaff cost approximately $13 million in suppression and postfire rehabilitation costs. However, a month after the fire a rare storm dropped two inches of rain in one hour over the burned area creating massive downstream flooding and the loss of one life. Homes were inundated, the pipeline for the municipal water supply of Flagstaff was damaged, and utilities were unearthed. The full cost of the fire as of March 2013 was over $133 million. In hindsight, it would have cost less than $5 million to treat a sufficient amount of the area to significantly reduce the severe fire behavior. The Schultz Fire provides a graphic example of how an investment in prevention and avoidance of catastrophic fire is justification unto itself for restoring forests.

Despite the jostling for position by the diverse stakeholders in the 4FRI, everyone is bound by the same ideals and *ethical framework.* "The Path Forward," one of the foundational documents of the group, states, "We expect that landscape-scale restoration across the Mogollon Rim will support healthy, diverse stands, supporting abundant populations of native plants and animals; thriving communities in forested landscapes that pose little threat of destructive wildfire; and sustainable forest industries that strengthen local economies while conserving natural resources and aesthetic values." The prize is clear; however, the vision can be lost in a struggle over details. We can only hope that, as more acres are restored and fire is tamed into a natural role, balancing the demands of people with the needs of ecosystems will become easier.

Closing Remarks

It is our opinion that ecological restoration is a management action that bridges the gap between preservation and conservation. Ecological restoration, by setting ecosystems on trajectories that lead to the recovery of more nearly natural, self-regulating ecosystem structure and function, can simultaneously protect nature while providing for sustainable uses by current and future generations. Such sustainable and adaptive ecosystem dynamics are especially critical in the face of climate change and population growth.

The current crisis in western forests demands action now;

there is sufficient information to design and implement scientifically valid approaches that will help determine how best to proceed. Sound knowledge to guide the restoration of frequent fire forests is available, based on the abundant scientific research that began in the 1890s and continues today. We have solid information about forest conditions before European settlement, changes in fire regimes, deterioration of overall ecosystem health, and ecological responses to thinning and prescribed burning—the key elements of any attempt to restore ecosystem health in ponderosa pine and related ecosystems.

Unless something is done to reverse the deterioration of ecosystem health, current and future generations will continue to incur increasing costs while simultaneously enjoying fewer benefits from public lands. It is increasingly clear that the costs to society are now so great that we can no longer afford to let these ecosystems continue on their current path. The risks of inaction far outweigh the risks of such treatments.

Preserving Nature on US Federal Lands

MANAGING CHANGE IN
THE CONTEXT OF CHANGE

Norman L. Christensen

The practice of nature preservation ultimately hinges on three questions. 1) What should we preserve? By this, I refer to the specific categories of things, as well as the items within those categories, that qualify as nature and that we deem worthy of preservation. 2) How should preserves be designed? We must determine how much, with what boundaries, and in what context preserves should be established. 3) How, exactly, should preservation be accomplished? What actions do we need to take to ensure restoration and/or conservation success? Answers to these three questions have changed considerably over the past century, and much of that change has come about as a consequence of changes in our understanding of change itself. Two events in the middle of that century mark an important transition in our approaches to nature preservation on public lands.

The first event was the 1963 release of the report of the Board on Wildlife Management in the National Parks chaired by A. Starker Leopold. Critique in the Leopold report, as it has come to be called, extended well beyond the management of large mammals to the general management of the parks' natural ecosystems. Perhaps the most iconic line in that report was the board's assertion that "a national park should represent a *vignette of primitive America*." As a primary goal, "the biotic associations within each park [should] be maintained, or where necessary recreated, as nearly as possible in the condition when the area was first visited by the white man." This language was completely consistent with the park mission as articulated in its 1916 organic act, "to conserve the scenery and the natural and historic objects and the wild life therein and to provide for the enjoyment of the same in such manner and by such means as will leave them unimpaired for the enjoyment of future generations."

The second event was passage of the Wilderness Act in 1964, arguably the first piece of legislation focused exclusively on the preservation of nature. This legislation created the Wilderness Preservation System representing a new category of federal land, wilderness, in which "the earth and its community of life are untrammeled by man, where man himself is a visitor who does not remain." Operationally, untrammeled meant no logging or mining, and only limited road construction. This was truly preserving nature.

Policies and practices for nature preservation prior to 1964 were applied with a confidence that bordered on hubris and based on firmly held beliefs about the roles of disturbance, change, and humans in nature. The answers to the three nature preservation questions were clear.

What should we restore and conserve? A vision of primitive America absent of the influences of European humans. This vision included the presence of relatively small populations of indigenous Native Americans as "noble savages" living in harmony with nature. It was also consistent with the then popular concept of the climatic climax ecosystem, unique self-perpetuating communities of organisms in equilibrium with the unique climates of particular regions. Importantly, these regional climates were themselves seen as unchanging over long spans of time. Large-scale disturbances such as wildfire, severe weather, and disease were generally understood to be localized and relatively unimportant.

How should we set boundaries for restoration and conservation areas? In ways that are politically and economically expedient. Discussion of the possible importance of spatial scale or context was virtually absent from the conservation literature of this era. Small-scale disturbances such as tree falls were believed to be responsible for regeneration of long-lived species, and most of the biodiversity associated with individual climax ecosystems was assumed to be contained within relatively small areas — thousands of acres, say. Policies and actions assumed that the preservation of nature at a place was dependent on actions at that place only; what happened on adjacent lands was of little concern. Thus, the boundaries of parks and national forests dedicated during this era were often either straight lines (e.g., the southern, western, and northern boundaries of Yellowstone) or they traced the channels of streams and rivers, thus dividing watersheds right down the middle. Nearby lands were logged, hunted, and developed as sprawling gateway communities as if these actions had no implications for the nature they surrounded.

How, exactly, should restoration and conservation be accomplished? Protect nature and let it be. Prevailing ecological theory argued that the best strategy for restoration of degraded lands and conservation of climax landscapes was protection from disturbance; natural change processes would then restore or preserve nature. Fire policy during this era was the most obvious and powerful manifestation of this strategy. Policies to suppress all wildfires on public lands were implemented early in the last century, and codified in 1930 in the so-called 10 a.m. Rule for fire management — "the aim for any wildland fire shall be to obtain control by 10 a.m. on the day after is first reported." Up to the 1960s there was general agreement on three points. First, fire has no natural role in most wildland ecosystems. Second, suppressing fire in wildlands would have no adverse consequences. Third, properly protected, ecosystems would change so as to create or preserve their climax characteristics and biodiversity.

The answers to these three questions indicate that nature preservation during this era was viewed as a form of museum curation. Nature preserves were the equivalent of "tree museums" where we preserved static "objects" rather than the dynamic processes upon which those "objects" depend. On degraded lands where restoration was needed, goals were (and often still are) articulated in terms of a "desired future state," a phrase consonant with the static vision for the nature of nature. Over the last five decades, we have learned that nearly all of the assumptions underpinning these answers were either simplistic or simply wrong. Our understanding of the role of disturbance and the ecosystem change it causes, of the importance of spatial scale and context on these processes and of humankind's place in nature has changed mightily.

Disturbance and ecosystem change. Some ecologists were questioning prevailing notions about natural disturbance and climax ecosystems long before 1964. In the 1920s, Aldo Leopold, then a Forest Service worker, cited evidence for the importance of frequent surface fires in arid forests of the Four Corners region in presettlement times and of the negative consequences of the loss of fire in these forests owing to cattle grazing and active fire suppression. With characteristic prescience, he argued that light burning was needed to restore healthy forest conditions. By 1964, ecologists were coming to understand that wildfire was a natural process in nearly all terrestrial ecosystems and that many species in these ecosystems were not simply resistant to and resilient from such fires, they were dependent on them. Succession following natural disturbances was not necessarily linear leading to stable climaxes; it was often cyclic—with cycles driven by ecosystem change that actually increased the likelihood of disturbance. Cycles might be short and punctuated by low-severity disturbance such as in grassland and savannas where low-severity fires burn every few years. Cycles might be longer and disturbances more severe such as in lodgepole pine forests where intense crown fires burn trees and accumulated fuels every few centuries. In any case, it was clear that suppression of fire in many of these ecosystems was resulting in changes that influenced the likelihood of future fires. In semiarid western shrublands and forests, fire suppression resulted in accumulations of flammable fuels that increased both the probability and severity of wildfires. In much of the East where moister conditions prevail, fire suppression produced comparatively nonflammable fuels and facilitated the invasion of fire resistant and intolerant species. Research over the past few decades has revealed that notions of regular cycles of disturbance and change are themselves simplistic. Prior to European settlement, many landscapes were mosaics composed of ecosystem patches that experienced highly variable disturbance intervals and disturbance severities. The biological diversity of such landscapes was dependent on the diversity and variability of its mosaic.

Spatial scale and context. The fact that the size of a preserve, its context, and the location of its boundaries influence conservation success has only been appreciated in the past few decades. The field of landscape ecology was born and has blossomed during this time, and the phrase "preserve design" has become part of the conservation lexicon. We now know that biological diversity is not a static property of an ecosystem; rather, it is the product of dynamic processes that influence the rate at which species immigrate to or disappear from it. That large preserves support larger, more stable species populations and, therefore, more species has become a core principle in ecology. This is even more important on dynamic landscapes where disturbance and change are responsible for much of that diversity. Ecologists define minimum dynamic areas (MDAs) for such landscapes as the territory required to capture the full range of disturbance and change to support biodiversity over long spans of time. MDAs are typically measured as millions and tens of millions of acres. That the character of boundaries and landscape context influence diversity of a place has become an equally important core principle. Landscape connections or their absence influence the movement of organisms across landscapes, and thus their ability to repopulate areas from which they have disappeared. These features also influence the spread of disturbances and the dynamics of landscapes.

Humankind's place in nature. The question of whether humans are part of or apart from nature is entirely academic. Imagining nature in our absence is useful only in the same sense that we might imagine movement through a frictionless universe. During the first half of the last century we vastly underestimated the influences of humans past, present, and future in Earth's ecosystems. In the millennia prior to European contact, Native Americans numbered in the tens of millions, and their activities profoundly shaped the nature we so much wish to preserve. They hunted hundreds of large mammals to extinction, and they managed the species that remained through the regular use of fire. The landscapes that greeted early North American explorers were very much a product of their activities; the details are not entirely clear but we know the nature and scale of those activities were changing through time. Human influences were even more pervasive as we began earnestly establishing and managing nature preserves in the first half of the twentieth century. While altering patterns of natural disturbance, we were creating novel kinds of disturbance. We were modifying landscapes in ways that limited the movement and accelerated the disappearance of native species, while facilitating the invasion and invasiveness of plant, animal, and disease species from other continents. We were mostly clueless about the changes wrought by our use of fossil energy for Earth's climate and the chemistry of its air and water, even as those changes had become ubiquitous. These human influences impact species directly; they also influence patterns of natural disturbance and the change processes they produce. We can no longer assume that natural disturbances and their associated patterns of change will preserve the nature we desire. There are now over seven billion of us; the most conservative estimates suggest that, 50 years from now, we will be striving to preserve nature in the context of two billion more of us, plying technologies we have yet to imagine.

This knowledge and these changes present formidable challenges as we now attempt to answer the key nature preservation questions.

What should we restore and conserve? Goals and strategies to preserve nature should include the full range in variation in species diversity and composition associated with disturbance and the change that proceeds from it. At scales from forest gaps to large landscape patches, much biodiversity is associated with disturbance and the succession that proceeds from it. Restoration and conservation of nature cannot be viewed as the curation of "natural and historic objects." Instead, we must define goals in terms of the dynamics of landscapes and the disturbances—the processes—that produce those dynamics. Restoration and conservation goals are often articulated in terms of "desired future condition" when they ought to be focused on "desired future change." We should care about history, but not too much. Historic range of variation in disturbance and the change it produces may inform restoration and conservation goals, but we live in a rapidly changing world. We cannot simply assume that restoring past processes will result in restoration or conservation success.

How should we set the boundaries for restoration and conservation areas? Pattern, scale, and context influence both disturbance and postdisturbance change, and preserve design really does matter. In this regard, we must worry as much about the territories we don't restore or conserve as we do about the territory explicitly targeted for preservation. In an ideal world, nature preserves would be scaled to their minimum dynamic areas that would encompass and stabilize changing natural mosaic. We have repeatedly learned over the past several decades that most of the landscapes that we have set aside for the preservation of nature are only a fraction of that size. Furthermore, the frequency and size of many disturbances, including fire and hurricanes, appear to be increasing in association with other human-caused change. Successful restoration and conservation strategies must be developed with the explicit understanding that managed territories are almost always smaller than the minimum dynamic area.

How, exactly, should restoration and conservation be accomplished? Restoration and conservation policy and practice in any ecosystem must be tailored to unique mechanisms and postdisturbance ecological legacies that determine the trajectory and tempo of successional change in that ecosystem. A half century ago, nature preserves were viewed as museums, and their curation was centered on protection from disturbance; restoring and maintaining natural disturbance regimes and key ecosystem processes are now the highest priorities. But ecosystem change following disturbance is far less predictable than formerly thought, even more so because of human-caused changes in climate, landscapes, and species distributions. Today, our attempts to simulate natural or past disturbance regimes or to encourage "natural" processes of change can have undesirable (unnatural) consequences such as the loss of biodiversity or the invasion of nonnative species. The world has never been the same twice, and in a world undergoing constant change, historical references are arbitrary.

All of this has significant implications for what it is we deem natural. In an ideal world, we would like to believe that we preserve nature when we preserve the natural processes that maintain it. Yet, we can no longer be sure that is true. We are now left with the unsettling task of defining nature for ourselves, and we must do this in the context of complex ecosystem change and a limited understanding of the consequences of our own actions. We are tinkering with nature, and we would do well to follow Aldo Leopold's admonition that "to keep every cog and wheel is the first precaution of intelligent tinkering." More than ever, preservation of nature demands humility, rather than hubris. Perhaps that ought to be our first precaution.

After Preservation—the Case of the Northern Spotted Owl

Jack Ward Thomas

As Norm Christensen (this volume) notes, Aldo Leopold famously opined that "to keep every cog and wheel is the first precaution of intelligent tinkering." That admonition came to fruition in the Endangered Species Act of 1973 (ESA). "The purposes of this Act are to provide a means whereby the ecosystems upon which endangered species and threatened may be conserved, to provide a program for the conservation of such... species." That would take place through emphasis on select — threatened or endangered species — whose welfare as assumed to reflect the health/welfare of the ecosystem of which they are part. As defined in the *Merriam-Webster Dictionary*, "To conserve is "to keep in a safe or sound state . . . *especially*: to avoid wasteful or destructive use." An ecosystem is "the complex of a community of organisms functioning as an ecological unit in nature." It must be assumed that the authors were precise in their choices of words and understood the difference between "conserved" and "preserved."

I was involved in three efforts—two in a leadership role—to comply with the ESA in the case of the northern spotted owl (*Strix occidentalis occidentalis*), or NSO, listed by the US Fish and Wildlife Service (USF&WS) in 1993 as "threatened" under the auspices of the ESA. These efforts culminated in the 1993 report of the Forest Ecosystem Management Assessment Team (FEMAT)—established by President William Clinton in 1993—entitled "Forest Ecosystem Management: An Ecological,

Economic, and Social Assessment" and jointly published by six federal agencies. The FEMAT's report also addressed the recovery of the marbled murrelet (*Brachyramphus marmoratum*) and coastal runs of salmon (*Salmo sp.*) relative to spawning habitats in old-growth forests. When President Clinton chose "Option 9" out of 10 options presented, it became commonly referred to as the "President's Plan" or, more formally, the "Northwest Forest Plan" (NWFP).

It should be noted that the USDA Forest Service (USFS) already had in place a policy (the "viability standard") that required managers of national forests to maintain habitats to insure viable populations of all native and desired nonnative vertebrate species. This self-imposed standard aimed at precluding listing of species as "threatened" or "endangered" under the ESA while maintaining the USFS's management autonomy. That approach proved—in an increasingly litigious atmosphere—impossible to meet given extant technical capabilities. It was one thing to set a standard and another thing to demonstrate compliance and efficacy. The "viability standard" was replaced in USFS planning regulations by a combination of "fine and course filters" to afford a less rigid standard of viability for species in question.

Most of the research that led to the listing of the NSO as "threatened" was conducted or sponsored by the USFS and provided the technical basis for the USF&WS's determination. In a review of the adequacy of the NWFP, a federal court in Seattle, Washington, required the government to provide an assessment of the impact on several hundred other species of vertebrates and plants that might be considered for listing as either threatened or endangered under the ESA. On the basis of an analysis by an interagency team of scientists, it was deemed likely that the NWFP would provide for the continued efficacy of those species. The federal judge accepted that expression of "expert opinion" and allowed the NWFP to proceed.

This approach to the "conservation of ecosystems" assumed that the viability of the old-growth ecosystems of the Pacific

Northwest (northern California, Oregon, and Washington west of the Cascade mountain range) could be "conserved" for the foreseeable future with forest management based on the hypothesis — intimated in the ESA — that the welfare and distribution of a chosen indicator species (i.e., "threatened or endangered" species) would be indicative of the welfare of all species (plant and animal) that make up the ecosystem in question. In this case, the management approach was based on the preservation/conservation of the "climax" stage of a forested ecosystem that had developed over many centuries under an array of climate regimes that occurred over many hundreds of years and are ongoing. It seems highly unlikely that "ecosystems" that do not replicate the processes and events that produced current old-growth conditions will produce replicas of current old-growth ecosystems from existing younger stands.

The management plan — that is, the "recovery plans" required by the ESA — for old-growth forest habitats was designed to sustain viable populations of three "indicator species" that had been declared, based on the preponderance of extant evidence, to be "threatened" by the USF&WS. Those decisions fulfilled the requirements of the ESA for a recovery plan for each of those species. It was assumed — in the ESA — that the welfare of those three subspecies would reflect the welfare for all other species (plant and animal) associated with that "ecosystem" in its current condition. In the case of the NWFP, that hypothesis has been ongoing for just over two decades.

During that time, an unanticipated event came into play. The barred owl (*Strix varia*), or BO, classified in the same genera as the NSO, is closely related enough to hybridize and produce reproductively viable offspring. That opens the question of how different the NSO and the BO really are. The BO expanded its range westward into the Pacific Northwest, theoretically as forest management in both Canada and the United States over the period of 1930 to the present produced more favorable habitat conditions. Upon arrival in the Pacific Northwest, that range expansion included remnant old-growth habitats occupied by the NSO. The timber harvesting, commonly using patch clear-cutting, coincidently altered old-growth stands favored by the NSO while producing more favorable habitat conditions for barred owls relative to northern spotted owls — including younger stands treated periodically by thinning and, then, re-peat harvesting at stand ages of 100 years or less. For example, stand composition and structure were much simplified in such circumstances and openings and younger stands were routinely available as habitat.

It became increasingly obvious that BOs were displacing NSOs in old-growth forests reserved from timber harvest under the NWFP to such an extent that the USF&WS instituted research testing of whether a program of controlling BO numbers to re-duce competition, predation, and crossbreeding would benefit NSOs.

This development brought the theory upon which the ESA was based — the welfare of threatened or endangered species as a reliable indicator of overall ecosystem health — into question. The reexamination of the foundational hypothesis underlying the ESA will beg the question of whether it is possible to "freeze" in place ecosystems that are in the process of evolving into an-other state that might well include a new mix of species — both plant and animal. That evolution, in turn, is likely to produce opportunities for species that evolved in forests with a closer resemblance to the stand structures and species composition of the "managed forests" of the Pacific Northwest, which have re-placed the original old-growth stands.

These changes resulting from "ecosystem preservation" raise a critical — and ultimately unavoidable — question. Is any approach to dealing with "ecosystem preservation" that depends main-taining a status quo — especially of a single "indicator species" — viable over the very long run? In this case, if the BO displaces and, to some unknown degree, fills the niche occupied by the NSO in the old-growth ecosystems of the Pacific Northwest, will that trigger further dramatic changes in that ecosystem? That seems unlikely. And, even if so, is this simply adaptation to changes in the forests of the Pacific Northwest and elsewhere— where some species are extirpated, some diminished, some enhanced, and so on?

If numbers of NSOs continue to decline, has the NWFP "failed" as an effort to conserve the old-growth ecosystem? If so, in what regard? Indeed, NSOs are becoming fewer—and may be destined for extirpation over much of their former ranges. Yet, to some unknown extent, the niche occupied by NSOs is being filled by another closely related species—the BO.

Is the "old-growth ecosystem" likely to "collapse" or even noticeably changed as a result? The answer could be hugely impactful in the social, legal, and economic senses. The extent of the impact will depend on how the USF&WS and the federal courts react to legal challenges relative to compliance with the ESA in the bellwether case of the NSO—the poster child of the ongoing "political brawl" over the approaches to ecosystem conservation. If those questions are narrowly decided by the courts, it could leave federal agencies little leeway beyond determined efforts at control of BOs. That would likely continue until the almost certain futility and costs—economic and political—of that approach becomes obvious. Or, conversely, until the huge gamble on the viability of a single subspecies as the primary indicator of overall welfare/function of an ecosystem has been brought into question and resolved.

Is the more significant question one of the overall impact on a threatened ecosystem by the displacement of a subspecies declared to be "threatened" under the ESA by a closely related species of the same genera in the process of increasing its range in response to changing habitat conditions? How different are BOs and NSOs in terms of their role and function in shared ecosystems? Is this displacement/replacement of one species by another species in the same genus triggering, or apt to trigger, a cascade of ecological consequences that will significantly impact the old-growth ecosystems of the Pacific Northwest? That seems unlikely and, even if so, that question can be investigated. Will we go on sending "barred owl control teams" into remnant old-growth stands? For how long? At what costs — economic and in terms of credibility? And how different really are NSOs and BOs in terms of their role in old-growth ecosystems?

If the "conservation" of old-growth ecosystems (as indicated by the welfare of the NSO) is the objective of the NWFP, the best that management agencies can do utilizing the management/legal processes prescribed in the ESA — guided by court rulings from time to time — is to buy time for adaptive/evolutionary processes to occur in the ecosystems deemed to be at risk. Such processes are ongoing — as they always have been and always will be. At present, there is no legal alternative in the United States — or credible ecosystem management alternative — to "conserving/preserving" old-growth ecosystems in the Pacific Northwest beyond setting set aside old-growth stands of appropriate sizes and arrangements on the landscape while encouraging/facilitating the development of replacement stands of large old trees. In doing so it can only be hoped that the young stands being managed with the intent of producing "functioning ecological old growth" sometime is the next century will fill that ecological role. Big old trees will not necessarily result in the old-growth ecosystems that they are intended to replace. In any case, the answer lies many decades — more likely centuries — into the future. Likely developing stands, and extant old-growth stands, will involve adjustments to evolving ecological conditions and will produce ecological communities to some unknown degree different from those that existed during the hundreds of years — even centuries — of evolutionary processes that produced the old-growth ecosystems we seek to protect today. And those ecosystems themselves are in transition.

About the best managers can do, given current knowledge, is to provide increasingly rare ecosystems opportunities to adapt to change through evolutionary processes. Anthropogenic in-

duced change in ecosystems will accelerate for the foreseeable future—even in those ecosystems in protected status. Indeed, "The times they are a-changin'"—as they always have and
always will. It has observed by Frank Egler that "ecosystems are
not only more complex than we think, they are more complex
than we can think." If so, it is well to be humble when we pon-
tificate about ecosystem preservation/conservation in an era of
likely accelerating climate change coupled with human population growth and demands on ecosystems to supply goods and
services.

In the case discussed here, the FEMAT assembled the 10 options and associated consequences to sustain viable populations
of the NSO as the surrogate for the old-growth ecosystem of the
Pacific Northwest. President William Clinton chose "Option 9"
to become the NWFP. The FEMAT was instructed, to the extent possible, to absorb the social/economic impacts on federal
lands. There is a lesson therein. In the case of the United States,
and likely elsewhere, the "best possible" preservation schemes
will, most likely, be limited to lands in public ownership. And, to
the extent that "preservation of ecosystems" constrains land uses
for other purposes, there will be significant social/political resistance. It is well to remember the old adage that "the most sensitive nerve in the human body runs between the heart and the
pocket book." The preservation/conservation of ecosystems of
which threatened/endangered species are part does not—and
will not—come "free for nothing." All lands managed for any
purpose—including conservation/preservation—come with associated direct and indirect costs. And, in such cases, how to
consider cost/benefit ratios is tricky at best. Given current eco-
nomic/social/political conditions in the United States today (in
2015), is it conceivable that (the circa) 1993 NWFP would stand
any chance of either development or adoption today?

As we live, learn, and evolve in our thinking of how to conserve ecosystem functions, it will be well to constantly upgrade

approaches to dealing with the "conservation of ecosystems" as a national policy. All of this leads to a critical political/economic/ecological question. Is the ESA — and the evolved processes required to comply with that landmark legislation — still the best approach to dealing with the conservation of threatened or endangered species as surrogates for the welfare of entire ecosystems that they represent and for which they serve as a poster child? Surely we have learned enough in the four decades since the passage of the ESA to evolve/devise better approaches to sustain ecosystems. We should recognize that they face both evolutionary pressures and processes. These demands will simultaneously accelerate and produce inevitable interactions with both climate change and attempts to meet the burgeoning demands of *Homo sapiens* resulting from population growth and increased demands for goods and services. There will be inevitable impacts on other species and the functionality of ecosystems.

Two decades of experiences with the NWFP — which relied on the status of species accorded threatened or endangered status to represent the welfare and functionality of an entire identifiable ecosystems — have revealed problems with the underlying logic. It is now more than four decades since the institution of the ESA in 1973, and we have, or should have, learned much in the course of our attempts to comply with adjustments required by new laws and subsequent court decisions. Is it not time to consider midcourse corrections?

It is well to understand that our relatively new concerns with "conservation/preservation" of threatened or endangered species and the ecosystems they represent runs contrary to the long-standing approach of getting more of what humans want/need/desire out of ecosystems. After all, it is through simplification of ecosystems that we have learned how to concentrate energy, nutrients, and water to magnify the production of desired products to support burgeoning populations of *Homo sapiens* who, beyond subsistence, seek a materially better life. In such circumstances, it is only realistic to realize that "preservation" as a management technique will be increasingly limited in the political, economic, and social senses as a viable alternative for informed management of ecosystems to be productive of goods and services increasingly needed/desired by *Homo sapiens*.

Celebrating and Shaping Nature

CONSERVATION IN A RAPIDLY
CHANGING WORLD

F. Stuart Chapin III

Nature has become an increasingly ambiguous concept in the rapidly changing, human-dominated world in which we now live. Nature might be idealized as wilderness—a pristine landscape or seascape without people, whose dynamics are shaped only by "natural" processes, unaffected by human actions. Alternatively, nature might be viewed as everything except built infrastructure, including urban parks and the lawns and channelized streams of a suburban subdivision. I suggest that both of these conceptualizations, and everything in between, are valid representations of nature and in many cases can motivate stewardship—the shaping of pathways of change to foster social and ecological resilience and well-being. It matters much less "what nature is" than what it means to us and how this meaning influences our behavior. Let's explore the limits of naturalness, and then think about how a spectrum of views of "nature" might influence people's attitudes about stewardship and sustainability.

Many traditional conservation programs have focused on the wilderness end of the spectrum, seeking to minimize the human footprint. Within the United States, Alaska, where I live, might be considered the epitome of wilderness. However, people have been an integral component of Alaskan ecosystems for at least 10,000 years, just as in other North American ecosystems prior

to Euro-American colonization. Both deliberate efforts to eliminate or displace Native Americans and inadvertent depopula-
tion through exposure to new diseases radically reduced Native
American population density. The settling of the West was, in
part, intended to fill and use these "empty" lands and later to set
aside some of them as wilderness "untrammeled by man, where
man himself is a visitor who does not remain," in the words of
the 1964 Wilderness Act. However, the emptiness of these lands
was largely an artifact of colonization. A more representative de-
scription of wilderness over the last 10,000 years would be lands
inhabited by people who depend on and interact with those
lands for food, shelter, and cultural identity. Wilderness management that builds a sense of human connection to these places,
rather than protecting them from people, would seem like the
most appropriate framework for stewardship. Prehistorically,
many of these wildlands had areas of very low population densi-
ties, where management to minimize human impact would still
be appropriate. Here the challenge is to foster a sense of place
and respect for nature that connects people to these lands in a
nondestructive way. This might be done by encouraging hiking,
canoeing, and widespread exposure to films that allow people to
experience and appreciate these lands respectfully with minimal
impact on the capacity of these lands to adjust to inevitable environmental changes such as changes in climate or disturbance
regime.

At the opposite extreme, people living in cities are surrounded
by built infrastructure that covers the land and invisibly channels
waters beneath impermeable surfaces. The built infrastructure is
the primary focus of people's activities, although lawns, parks,
cemeteries, and gardens, mostly dominated by exotic species
that are well adapted to urban environments, can be important
places for family activities, community interactions, or sources
of memories that tie people to the landscapes of their historical
roots and therefore to their sense of identity. Conservationists
largely ignore urban and suburban nature, which falls within the

domain of landscape architects who tend to treat it as an aesthetic or functional extension of the built environment, whose primary purpose is to serve societal needs. Here the challenge is to encourage the more regular engagement of people with these natural elements and to provide a complement to built infrastructure as a venue for life's experience. Urban gardens and parks, suburban trails, and stream cleanup projects provide opportunities for people to engage in urban nature.

Other parts of the United States are intermediate, with forests that regenerated after logging, agriculture, or mining and channelized streams and farm ponds that are later left to their own devices. The functioning of these ecosystems often has a strong legacy of their earlier human impacts, with a plow layer beneath a forest soil or a forest organic layer beneath a pond. In these places, species composition is likely to be a hybrid of many former ecosystem types, or the composition may be largely novel. Yet this is where most American kids fish and find frogs, families go on picnics, and lovers look for solitude — if they interact with nature at all. Much of the nature that surrounds the San Francisco Bay area, for example, is dominated by Australian *Eucalyptus* and European grasses, but it is highly valued by urban people seeking to spend time in nature. The challenge in these ecosystems of intermediate human impact is to celebrate their current ecological beauty and function and their current and future capacity to envelope people in nature. In terms of the area of nature in the United States and total number of people who are likely to experience and identify with nature, these may be the most important opportunities for stewardship. These areas provide opportunities for all people to interact with nature and allow us to learn from the huge number of human, environmental, and biotic "experiments" that have altered these landscapes and seascapes in the past and will continue to do so in the future. This provides many pathways to introduce or strengthen the ties between people and the rest of nature and to learn from these interactions.

Natural ecological processes, such as disturbance, competition, predation, nutrient cycling, and energy flow shape the species composition and functioning of ecological systems across the entire spectrum of naturalness, although with quite different controls and dynamics. In developed areas, there are islands of relatively unmodified ecosystems in a matrix of novelty. In areas with sparse human population, there are islands of human disturbance in a matrix of "wilderness." Across this spectrum the islands and matrix interact, so that each changes the other. Superimposed on these novel landscape mosaics are changes in climate and development, which increase the likelihood of species introduction and movement and therefore the interactions between island and matrix. These dynamics suggest that no ecosystem is likely to maintain its historical structure or function and that those ecosystems with which people are most familiar are likely to have changed substantially from their historical analogues. At the same time, the dynamics of ecosystems across this spectrum are products of fundamental ecological and cultural processes and are therefore a potential source of understanding and of connection of people with the rest of nature.

Given the strong human signal in all ecosystems, what actions can be taken to foster a more sustainable future, and how might this vary across a spectrum of naturalness? People's interactions with nature are influenced by interactions among what we are allowed to do (laws and regulations), self-interest (e.g., balance of economic costs and benefits), and what we want to do.

For the most part, laws and regulations constrain unsustainable behavior that might otherwise occur when people or businesses seek to maximize short-term individual benefits and leave society to cover the environmental costs. Regulations that restrict the release of pollutants to air or water, for example, protect these public goods from the actions of individual polluters. Similarly, zoning restrictions on development in areas with scenic or wilderness value protect these areas for society as a whole, both today and in the future. Regulations are generally important but insufficient for the stewardship that shapes trajectories toward a more sustainable future.

Economic incentives often prioritize short-term benefits to individuals despite longer-term costs to society. People often choose to purchase, for example, products that cost less but have greater environmental impacts than more costly, environmentally friendly products, or voters may favor developments that provide jobs over the protection of an area to preserve its natural qualities. Alternatively, conservation and agricultural easements that reduce taxes in rural areas with rising property values allow landowners to maintain indefinitely those qualities of the land that contributed to their own sense of identity and connection to nature. Quantitative assessment of ecosystem services is another approach to incorporating natural values in trade-off calculations (e.g., provision of clean water by forests, reduced spread of pests across ecological buffers, storage of carbon in forests). The market exerts a powerful influence over choices that people make, so an ecosystem services framework that incorporates the economic value of these services in land-use decisions increases the likelihood of environmentally friendly decisions.

People often interact with nature because they want to, whether this is to barbeque on their lawn, weed their garden, walk in the park, purchase a home in a pretty place, go camping in the mountains, kayak in a remote fjord, or derive satisfaction from habitat protection for an endangered species that they will never see. Although these cultural or aesthetic values can be incorporated into an ecosystem-services framework, they are more simply viewed as a sense of connection to nature. Sense of place is "the collection of meanings, beliefs, symbols, values, and feelings that individuals and groups associate with a particular locality." Sense of place can contribute to the sense of identity of individuals and groups associated with a place, build attachment to that place, and help frame environmental debates surrounding that place. Places that people care about can create a spectrum of potential responses to resource issues, as often seen in debates over conservation versus development among people who feel strongly about a particular place. The challenge in using sense of place as a motivation for stewardship actions is to draw on those attributes that unite people in their commitment to place and to negotiate acceptable solutions to those political, ideological, and social issues that are contested and create conflicts about how people should interact with these places.

If a broadly shared sense of place can be an important motivation for stewardship, how can it be fostered, and how should the approach vary across a gradient of human modification? In general, shared experiences, stories, or education that strengthens personal and cultural attachment to place can build a sense of connection. For example, a community garden in a vacant lot or a school project to clean up a polluted stream can turn areas of neighborhood hazard and avoidance to a source of community cohesion. At the less human-modified end of the spectrum, actions that identify vulnerabilities or reduce threats (e.g., removal of invasive species, monitoring of endangered species, or zoning to restrict development) may provide a similar sense of connection and commitment to particular places.

I suggest that nature should be celebrated across a broad spectrum of naturalness and that landscapes be conserved and protected across this spectrum in ways that foster human connection and identity with the nature that each provides. The rapid urbanization that is occurring globally and in the United States provides opportunities for building these connections in both rural and urban areas.

Population declines in rural areas provide opportunities for protection of the ecosystems that occur there and the opportunities for people to interact with and participate in nature in these places. Conservation easements, tax incentives that enable ranchers to maintain "working landscapes," and zoning that focuses development in some corridors and protects wildlands in other areas provide opportunities for people to connect with (relatively) wild nature, while protecting people's homes and other infrastructure from wildland fires.

Urbanization generally involves replacing or building new infrastructure every several decades. Most of the urban infrastructure of 2050 is yet to be built! These redevelopment actions are opportunities to allow urban nature to develop and flourish in a landscape that fosters regular interaction with the people that inhabit this land—as forests and grasslands, parks, gardens, and other open spaces. These places are particularly important because they constitute the nature with which most people will interact.

In both rural and urban areas it is important that people interact with nature in ways that improve their understanding of how nature responds to human and other disturbances. These activities might include building birdhouses, fishing, gardening, farming, participating in prairie or stream restoration, or any of a myriad of interactions. Only by becoming part of nature, through recognizing how nature affects us and how we affect nature, can we identify with nature and move beyond the notion of nature as a museum that has no direct connection to our lives.

Move Over Grizzly Adams— Conservation for the Rest of Us

Michelle Marvier and Hazel Wong

Conservation organizations do a lot of hand-wringing about the lack of diversity in their leadership, staff, membership, and donor pools, as well as the lack of gender diversity in their leadership ranks. The concern, of course, is that if conservation organizations appeal primarily to whites and the white proportion of the population is declining, then support for these organizations is destined to similarly decline.

Data from the US census provide some context to this problem: As of 2012, women are 51 percent of the US population. Blacks, Hispanics, Asians, and other minority groups constitute 37 percent—a proportion expected to exceed 50 percent by 2042. Additionally, almost 13 percent of the US population is foreign-born, and about 20 percent speaks a language other than English at home.

In stark contrast to the diversity of the United States, fewer than 18 percent of the scientists in the US Department of Interior belong to minority groups, and only one-quarter are women. People of color comprise only 12 percent of the staffing base and less than 5 percent of the board members of nongovernmental conservation organizations. Indeed, the American conservation movement has long been dominated by white males, and much has been written about why the green movement has remained so white. It's not just nature-related careers that lack diverse faces. The National Park Service has long struggled to attract more African Americans to our national parks. Hiking and backpacking are less appealing activities for many nonwhite groups, which sets up a flawed interpretation of who values nature in America: in the view of some nature lovers, picnicking is not on par with skiing and backpacking.

The underrepresentation of diverse faces in America's natural spaces and of women and people of color within the leadership ranks of the organizations that protect our nation's natural resources has an ironic twist—public opinion polls consistently find that people of color and women express greater concern for nature and the environment than do their white male counterparts. In fact, polls show that persons of color and women (groups that are overrepresented in the lower socioeconomic strata) are more likely to vote in favor of conservation initiatives, even when the initiative includes a tax increase that will directly impact them. In short, protecting America's lands and waters is a value shared by many, yet for decades conservation organizations have systematically neglected outreach beyond their current base of white donors and members.

An important question then is this—if people of color and women are more likely to express support for conservation in opinion polls and at the voting booth, why are they less likely to be members, donors, and leaders in conservation agencies and NGOs? We think the problem largely lies in the disconnect between the way professional conservationists have traditionally framed themselves and their mission and the reasons that underlie women's and people of color's support for conservation. As we develop these arguments, we also consider how conservation organizations might motivate a broader base of public support in an increasingly diverse America.

What's Not Working with How Conservationists Have Traditionally Framed Themselves and Their Mission?

In case you didn't grow up in the 1970s, the Grizzly Adams we refer to in the title of our essay is a fictional character (very loosely based on the life of James Capen Adams) who appeared first in a book, then a popular TV series and movie. The basic plotline goes like this: Grizzly Adams, wrongly accused of murder, flees into the wilderness where he fashions all the tools he needs to keep himself fed and healthy, builds himself a log cabin, and discovers that he possesses a special ability to befriend animals. This image of a white male, out of place in human society but at harmony with wild nature, could be dismissed as cheesy 1970s TV tripe. Certainly we do not hold a single TV show accountable for the marginalization of minorities and women from the conservation movement. However, it is the case that popular culture both reflects and reinforces societal images of who fits into certain roles and who is left out.

As African American author and conservationist Audrey Peterman noted, "The fact that we don't see ourselves in images on TV or in the outdoor media dampens our presence in this movement." And once you start looking for it, the predominance of Caucasians in American images of nature is striking. Carolyn Finney at the University of California, Berkeley, scanned 44 issues of *Outside* magazine from 1991 to 2001 and found that of more than 4,600 photos of people in the magazine, only 103 were of African Americans.

It is our personal observation that the Grizzly Adams trope—a white man, alone, self-reliant, and completely at peace in the wild—is a fantastical identity widely held within the conservation community. The stories of Henry David Thoreau, John Muir, and Edward Abbey, with their focus on the value of solitude in wild nature as a way to heal the spiritual damage of modern society, bear some similarity to Grizzly Adams's tale. Indeed, the identity of the white male toughing it out solo in America's great outdoors has roots in the historical legacy of who could own land in America. But it is time to modernize this self-image, which can make others feel excluded from the conservation movement. Not doing so will have severe implications on our future ability to protect our precious natural resources.

That's because in our pluralistic society, the voice of constituencies plays a key role in the policy-making process, and more vociferous and diverse voices are needed now and in the future to advocate on behalf of America's lands and waters.

The Grizzly Adams types within the conservation movement must make room for new faces. The challenge to all of us is to reimagine our definition of conservationist and nature lover. We need images of hikers in hijab, urban Hispanic women restoring rivers, African Americans tagging arctic birds, Asian Americans protecting wetlands, people of all races and origins loving and caring for nature. We need concerted efforts to diversify the faces we see in conservation organizations and in the outdoors. But, just as importantly, we need to frame our messages and motivations for protecting nature in ways that speak to the values, not just of the nation's Grizzly Adams types, but of diverse groups of American men and women.

Disseminating cultural images of minorities and women of all backgrounds enjoying nature and outdoor recreation is one piece of the puzzle. For example, the African American Environmentalists Association produces a TV show called *Urban American Outdoors*, groundbreaking in its efforts to showcase diverse people both working and recreating in the outdoors. Simply seeing someone who looks like you engaging in these activities can make them seem more doable, but facilitating the engagement of diverse people in outdoor activities may be an even more effective strategy. Much of the challenge is simply overcoming the initial barrier of never having tried it. Many groups are working to build stronger connections between minorities and wild nature. Rue Mapp of Outdoor Afro arranges birding trips for urban blacks. The Boy Scouts of America and the Girl Scouts are working to increase the diversity of their memberships. Positive immersion experiences, especially for children and teens, can lead to a lifetime of outdoor recreation, of nature loving, and ultimately of increased support for conservation.

More than just positive recreational experiences will be needed

to create future conservation professionals. One effort that
moves beyond mere nature immersion is The Nature Conservancy's LEAF program, which stands for Leaders in Environmental Action for the Future. This program provides paid in-
ternships for urban high school students, mostly students of
color, to work outdoors all summer on a conservation or restoration project under the mentorship of a conservation professional. Assessment of LEAF participants reveals the positive influences this program has not only on the interns' environmental
knowledge but also on their self-confidence and self-reliance, an
effect especially pronounced among girls. Alumni surveys indicate that these students pursue nature-related careers at a rate
that is nearly six times higher than the national average. Inter-
estingly, the assessment of the LEAF program focused not only
on the students, but also examined impacts on the conservation professionals who mentored the students. The assessment
found that participating in the LEAF program improved the cultural competency of both the students and the conservation professionals. The program influenced practicing professionals to
think differently about who counts as a conservationist as well
as the diverse reasons people may come to care about conservation.

   As we mentioned earlier, it's not just the diversity of faces that
needs changing in conservation. It's also the framing of conservation's mission and the rationales we provide to motivate conservation concern and action. The traditional focus has been on
remote wilderness and the protection of nature for its intrinsic
value, or the inherent right of diverse species to exist. These are
important goals, but a myopic focus on wilderness, remote from
cities, hinders broad support for conservation. The US Census
found that as of 2012 more than 70 percent of the US population
lives in cities of 50,000 or more people, and this percentage is
expected to grow. Conservation cannot focus solely on remote
wilderness accessible only to those with enough leisure time and
money for travel.

When wilderness areas are indeed the focus, conservationists must better communicate to all people why protecting those lands and waters benefits society. We must encourage people to visit their state and national parks (as intensity of support to protect the environment correlates with time spent outdoors), but the conservation tent must also be open to those who may not want, or cannot afford, to go backpacking or skiing. Recently, certain conservationists such as Dave Foreman have made appeals to "true conservationists" and "wild lovers" to "take back conservation." We find such discussion of "true conservationists" reminiscent of divisive rhetoric regarding what constitutes a "true American" or "true Christian." There is more than one right way to engage with nature, and it is counterproductive to build a wall between die-hard conservationists who relish week-long backpacking experiences and others who would never by choice go a day without a hot shower.

Undoubtedly, protection of relatively wild places is important, but the movement is stagnant, as illustrated by a sharp decline in people identifying as environmentalists and conservationists. The only way to reenergize it is to ensure broader participation. It is important that as many people as possible feel a connection to nature, not just those who can and want to access remote places. Hispanics, for example, tend to use parks extensively, but more often they visit outdoor spaces nearer to home and for group outings, rather than venture solo into the woods to be "one with nature." Conservation is not just about wilderness and wild lovers. There are many diverse ways to enjoy, love, and protect nature.

Engaging people with nature and encouraging conservation support in any way is a victory. Declaring love of nature to be an all or none proposition will only serve to discourage mainstream support and hurt the cause. The Nature Conservancy-sponsored focus groups in the Gulf Coast states, for example, recently found that just the mere mention of climate change can eliminate support for marshland and oyster reef restoration. This is an unintended consequence of the hardline approach, and can actually do a disservice for conservation. Drawing lines in the sand and failing to negotiate may feel self-satisfyingly heroic, but it is also ineffective. Finding ways to match conservation messages to the values diverse people hold is what will help conservation succeed.

Large national surveys consistently find that a majority of respondents in all categories feel that the benefits nature provides to human society (including marketable products such as fish, crops, timber, and medicinal ingredients, as well as benefits such as air and water purification by trees, and hurricane/flood mitigation by wetlands) are extremely important. Voters of color were far more likely than white voters to view these benefits to people as extremely important. These findings are consistent with data from the United States showing greater concern among Hispanics and African Americans, compared to non-Hispanic whites, about environmental matters directly affecting quality of life in their communities. Of course, we are discussing broad trends here. Can one find a black woman who cares more about biodiversity or wilderness experiences than clean water? Of course. And are there white men who care more about storm protection than about the transcendent beauty of a protected coastline? Yes. But, on average, messages that highlight the benefits of conservation for people are more effective with a broader constituency.

The power of demonstrating and marketing the benefits of conservation for people are now widely appreciated, as evidenced by the fact that, to some degree, nearly all major conservation organizations have embraced conservation for nature's benefits. The Conservation Campaigns Team at The Nature Conservancy has generated over $49 billion via state and local ballot initiatives because they highlight the benefits of nature to voters in all their voter outreach campaigns. Yet, some conservationists have not just resisted this trend, but have aggressively attacked a more human-centric framing. There seems to be angst that broadening the diversity of motivations and messages for conservation translates to a call for conservationists to abandon the argument that we should protect nature for its own sake. This is not the case. The intrinsic value argument has its merits and certainly appeals to some. However, nature for nature's sake does not attract enough people, or a broad enough segment of the population, to succeed on its own. Both arguments—protecting nature for nature's sake and protecting nature for people—can and should coexist. Using them together, rather than distinctly, will broaden the support base for conservation.

The American conservation movement's many successes—the nation's majestic national and state park systems, as well as the vast lands protected by private trusts—are testament to conservation's visionaries as well as its worker bees. The US Endangered Species Act of 1973 continues to serve as a model for conservation legislation in nations the world over. But the United States is undergoing enormous and rapid demographic, economic, and social and technological changes. Strategies that worked in the past won't work in the future. Conservation needs to stay ahead of the change and be prepared to lead for the future. We remain optimistic that we can and will save great portions of nature, even as the population, diversity, and economic wealth of Americans increase. But achieving this vision will require that we promote a more inclusive vision of who counts as a conservationist and embrace a more diverse set of motivations for caring about nature.

Endangered Species Conservation

THEN AND NOW

Jamie Rappaport Clark

Rachel Carson and Aldo Leopold have been heroes, friends, and mentors to me throughout my wildlife conservation career. I know it must seem odd to say that people I have never actually met are friends and mentors, but that's how it feels.

I was first introduced to Rachel Carson in high school. Her book *Silent Spring* spoke to me like nothing I had ever read before. As my own career path traced hers, first as a biologist, then working at the US Fish and Wildlife Service, and now at Defenders of Wildlife, her words of caution have continued to guide me.

Aldo Leopold was also an icon of mine as I studied wildlife biology in college and graduate school. Not only did his poetic and profound words ring true but so did his environmental land ethic and his understanding that the linkage between people and the land is common sense, yet so rarely considered.

If only I had the poetic abilities of Leopold or Carson—to convey the importance of nature to the American people in such an inspiring way that it would shake up our elected officials, mobilize a frequently disengaged public, and inspire opinion leaders. If Carson and Leopold were here today, would they still be able to move people to action as they did decades ago with the profoundness of mere words, or would they have to rely on savvy public relations firms, cool viral videos, and catchy slogans to get their points across?

While the modes of communication may have changed drastically since their days, the struggles we face have not. Their words are as relevant today as they were back then. The challenges they wrote about, including the deadly impacts of pesticides and pollution on our lands, water, and wildlife; the destruction of wilderness; the threat of uncontrolled development; the balance of nature and the debate over the role of predators in a healthy environment are as relevant now as during their times. A student today, unfamiliar with their writings, would never suspect they were reading books published a lifetime ago.

As I reflect on their works and reread various passages of their books, I find that neither of them was a strict preservationist in the purist sense of the word. They were pragmatic conservationists. They understood the needs of people to provide for themselves and their communities. But they also cautioned that the instant results we think we want might not be beneficial to us in the long run. They focused on long-term consequences of overuse and a lack of respect for the landscape.

Leopold advocated for a balanced approach to land management, stating, "The art of land doctoring is being practiced with vigor, but the science of land health is yet to be born." Rachel Carson furthered this approach. She did not advocate for the banning of all pesticides, but she did advise using as little as possible to limit the development of resistance. Her combination of compelling scientific evidence and riveting prose helped politicians and the public understand that all species, including people, are connected in a fragile ecological web, and that harm to one creature results in harm to us all. Finding that balance and respecting the intricacies of nature was their overarching conservation message.

Historically, we have put places aside for the preservation of wildlife and landscapes. But that by itself does not lead to true balance and, in the end, will not fully benefit nature or ourselves. Nature cannot thrive with only pockets of protection. Wildlife needs room to roam, the natural world has no boundaries; it's all around us. It's time we appreciate that we are part of our own environment, part of the web of life, and what we do to our planet impacts our own lives and well-being. Ultimately, the future of this life-supporting planet of ours is dependent on our moral and ethical values and a deeper understanding of the intricate relationship we have with the natural world.

We have taken on some extraordinary environmental challenges over the past four decades, but still struggle with the balance between the needs of nature and the needs of people; and the common philosophy seems to be that these needs compete. It is essential that we universally come to the same realization that Carson and Leopold did many decades ago that our fate and that of nature is intertwined and inseparable. Only then will we make consistently smart decisions that will benefit the whole.

One of our biggest challenges today is our expanding population. Back in Carson and Leopold's day, nearly three billion people (1950 figures) inhabited the planet. Today, there are seven billion people, and that level is expected to rise to 10.5 billion by 2050. This dramatic increase in our population places a heavy demand on the earth's resources: more food needs to be grown and processed, more clean water needs to be available, and more energy needs to be produced. And all of these people place a heavy strain on our environment: producing more waste that needs to be processed and disposed of safely and pollution that fouls our waters, lands, and air. The results are habitat loss and fragmentation, aggressive energy development and resource extraction, excessive pollution, invasive species, and overharvesting, which all wreak havoc on our planet.

We are already struggling with the impacts of the strain that industrialized and emerging countries have placed on the earth. Climate change, brought on by the emission of greenhouse gases, is causing our planet to warm, leading to extreme weather events and disrupting growing seasons that threaten our communities, our wildlife, and the habitat they need to survive.

The impacts on wildlife are the proverbial "canary in the coal mine" for us all. Elevated water temperatures can lead to deadly results for fish such as salmon, trout, and others that we rely on for food. More storms, floods, droughts, and fire damage habitat for wildlife and have devastating impacts on communities. Rising sea levels flood coastal communities and areas where vulnerable species like sea turtles and shorebirds live. Many species will lose their habitats completely, and that plight is not just reserved for wildlife. Already we are seeing communities that have been repeatedly pounded by storms, fires, and drought. It's time to take a step back and acknowledge that we all have an ethical responsibility to be good stewards of the planet, to conserve all native species, to maintain the life-support functions of natural ecosystems, and to protect for future generations the maximum benefits inherent in a rich diversity of species.

Saving the environment and our wildlife is not a luxury; it is a prerequisite to our own survival. Our needs are no different than theirs: we all need clean air and water and a place to live that provides us with sustenance. We all have a stake in this; and if we take steps that allow us to coexist with wildlife, we will be moving in the right direction. The challenges we face today are daunting but not insurmountable. I think the biggest challenge is coming to an agreement that we have to address the issues at hand.

Here at Defenders of Wildlife, we believe that our future depends on coexistence with our natural world. The vast majority of our land is already dominated by human activity, either directly or indirectly, so we focus on the conservation of native species in their natural communities with a priority on where people and wildlife meet.

First and foremost, we're concerned with preventing species and unique habitats from going extinct. All species have a right to exist. Other authors in this book have noted Leopold's indelible aphorism, "To keep every cog and wheel is the first precaution of intelligent tinkering." Too many animals and plants are

Figure 19. Our fate and that of nature is intertwined and inseparable. An ethic of coexistence in a changing world reaffirms the values of conservation and our sense of responsibility to future generations who deserve a healthy thriving environment. Photo credit: Hank Perry.

already in serious trouble, and it is our duty as responsible stewards to do everything we can to save them. Sure, we will have to prioritize what we take on first, but we cannot simply sit by and watch as more and more species go extinct.

Next, we must protect those species with an uncertain future. Climate change, habitat loss, development, overharvesting, and the impacts of invasive species place a heavy burden on the survival of many species in the near future. We must try to ensure that our actions do not lead to adding even more species to the endangered list.

And finally, we must restore important landscapes that are critical for wildlife, we must find ways to live more sustainably on the landscape, and we must deploy technologies smartly to enhance biological diversity. We need to learn how to better coexist with wildlife.

Coexistence can be on a small or large scale. At Defenders,

we have worked with ranchers to provide them with the tools they need to ranch safely and successfully in wolf and grizzly bear country. We have worked with small farmers to give them the knowledge they need to protect their livestock from endangered panthers in Florida. And we have provided guidance to hikers and others on how to live and recreate in bear country. But as we continue to put more pressure on our planet, we need to consider even broader ways to coexist.

By protecting beaches and wetlands from erosion, we are protecting habitat for imperiled nesting shorebirds and sea turtles, many species of fish, crabs, shrimp, and waterfowl. These natural areas buffer the coastline from storms, a critical defense line for shoreline communities.

By ending the use of lead ammunition, we protect wildlife such as endangered California condors from dying of lead poisoning. We have already taken lead out of paint, gas, and pipes, and it is no longer used in waterfowl hunting. Lead is not good for anyone, animals or people; so by not using lead ammunition, we will be saving both wildlife and people from this deadly toxin.

By providing safe passage across busy roadways with tunnels or overpasses, we have protected migratory corridors for many species of wildlife including Florida panthers, bears, caribou, and others.

By protecting bees, bats, butterflies, and other pollinators, we will strengthen our food supply. Chemicals that we use to control pests also destroy these valuable pollinators. By poisoning them, we harm ourselves. By protecting freshwater mussels from pollution, we protect critical water supplies essential to wildlife and thriving communities.

By advancing energy sources that are not as polluting as oil and gas, we will decrease the amount of greenhouse gasses going into the atmosphere. But we need to be smart about this energy transition and not jump blindly into renewable energy without thinking about the consequences of where it is located. Wind and solar energy can impact wildlife: birds and bats are killed by the blades of wind turbines, acres of solar panels carelessly placed across a biologically sensitive landscape can destroy important habitat for a wide variety of imperiled species, and some renewable energy systems use excessive amounts of water, depleting already stressed aquifers. But if we place these systems in the right low-conflict locations, we can coexist: providing safe, renewable energy for our future and protecting habitat necessary for imperiled wildlife and people to survive. We have to sit down at the table with all interested stakeholders and develop solutions that conserve our nation's biological diversity. We all must be leaders.

It's also time for those in political power, who have put short-term profits over long-term survival and common sense, to change their priorities as well. As President Obama stated in a speech on climate change, "We do not have time for a meeting of the flat earth society." It's time to reaffirm the values of conserving nature and our sense of responsibility to future generations who deserve a healthy thriving environment.

There are great challenges ahead of us. And we have already seen our future: record droughts, intense storms, oil spills, invasive species, and vicious wildfires. Forty years ago, we were at a crossroads: our air was dirty; our rivers were so polluted one even caught fire; and wildlife was going extinct. We were facing a stark choice. Would we as a nation accept a future of vanishing species and dirty air and water, or would we have the wisdom and courage to protect our nation's wildlife and natural resources for the future? Thankfully, we chose protection and conservation, and passed the Clean Water Act, the Clean Air Act, and the Endangered Species Act. These have been great tools that have protected our environment for generations.

Climate change, the demands of a growing population, and diminishing political will are placing us at a crossroads again. We all must come together. We cannot sit back and wait for other countries or communities to take the first steps. We must lead and others will follow.

Leopold was right in that conservationists are no wiser than others; we are just more appreciative of the important role that nature plays in our lives. Leopold and Carson were gifted with the ability to write in ways that moved people to care about the environment. And through their words, they educated generations to think responsibly about the land, to understand not only our power to destroy natural resources but also to nurture them. It's time we rattle the cages again and illustrate to the American public a clearer view of the importance of the life-giving resources this planet has to offer. Rachel Carson said it best, "man is a part of nature, and his war against nature is inevitably a war against himself." It's time to rededicate ourselves to saving each other.

Resembling the Cosmic Rhythms

THE EVOLUTION OF NATURE AND STEWARDSHIP IN THE AGE OF HUMANS

Amy Seidl

The document box delivered to me at the broad, oak desk was unadorned, disguising the treasure inside. Of the set of Rachel Carson's archives stored in the Bienieke Library at Yale University, I had chosen one box to spend my day with, the one labeled "Sense of Wonder: Writings, Edits, and Correspondence." Inside the box stood a dozen manila folders each carrying a loose sheaf of notes and clippings. Opening the folders, I entered a near-sacred space; Rachel Carson, the writer, biologist, and steward of American nature, became my intimate, and I moved into her circle of confidence.

Pages of notes spilled onto the silent table. Many were crafted in a woman's curved penmanship where the tip of a sharpened pencil was still evident. Pouring over the contents, I found articles clipped from conservation journals and newspapers with Carson's slanted marginalia to the side. One article described the ecology of a wetland and how it was home to painted turtles. Another recorded how the eggs of bald eagles were shattering beneath the bird's feathered, muscly legs due to the effects of DDT, the pesticide Carson condemned in her book *Silent Spring*.

There were letters in the folders too. Editors and publishers wrote to praise her lyrical prose and accept her articles. Skeptics requested more explanation for her claims or outright doubted her interpretation of the science. But the majority of letters were written by admirers who shared with Carson the environmental dilemmas from their own backyards and how they too felt the call of stewardship, not only at home but also in the parks and schoolyards they shared with their neighbors.

It is well known that Rachel Carson received both praise and abject criticism after publishing *Silent Spring*, the book she reluctantly turned her attention to after beginning what would posthumously be published as *Sense of Wonder*. The affection she displayed for the sea and its coasts was replaced in *Silent Spring* with the voice of Cassandra and a plea that Americans see how blunt and brutal industrial chemicals were to the unprotected constituents in nature. If, in her first books, Carson wooed us, deepening our love for life's mystery, in *Silent Spring* she moved us from awed observer to vigilant defender, asking that we steward the American landscape from invisible yet insidious threats.

Critics responded to Carson's polemic in several ways. The most base among them attacked her person, naming her a lunatic and "soft-headed." But there were two other camps of critics whose views inform nature preservation today. One assumed a vigorous, anthropocentric stance and responded to Carson's claims with an uncontained hubris, confirming that humans were *meant* to control nature, indeed that was our ecological role.

A second camp echoed elements of Carson's own empathy with the vastness of the natural world. Ironically though, these critics concluded that nature was far too big, far too powerful, and far too dominant as compared to humans to even entertain the idea that we could "manage" a planet. Humans could no more pilot Earth's course than we could steer its revolution around the sun or alter the salinity of the sea.

Fifty years after the publication of *Silent Spring*, the debate among Carson's critics has bearing on preservation in the Age of Humans for there's truth in both arguments: our actions, as it turns out, are powerful enough to compete with and at times dominate nonhuman drivers in nature. Moreover, humans now affect the course of nature and its very fabric: the evolution of species over time.

When the Age of Humans began is debatable. While not explicitly referring to the Anthropocene, nature scholars like George Perkins Marsh were potent spokespersons for how Americans were radically changing the environment. While American landscapes were significantly refashioned during the Industrial Age and resources were extracted with great speed — forests became agricultural lands and then forests again in a matter of a century — we viewed the visible changes in our landscape as solely ecological and began a movement to preserve their ecological function. But how many early preservationists anticipated their role to include the stewardship of evolutionary outcomes as well as ecological ones? Moreover, how aware would the next wave of preservationists be to the fact that humans could drive genetic change, fathoming that our actions are analogous to the clearly nonhuman forces of plate tectonics, glacial recession, or island formation that were responsible for the evolution of life documented in the fossil record? But this is what we've found: humans influence Earth's biotic communities the way the geometry of the solar system does. By acknowledging our wholesale influence, including the fact that anthropogenic climate change has eclipsed the next glacial age, contemporary stewardship is presented with a Gordian knot; humans are both the protector and the instigator of the evolution of life on Earth.

*   *   *

When I was a child growing up on the high plains near Cheyenne, Wyoming, my parents would gather with fellow preservationists in our living room, a group called Wyoming's Outdoor Council. There they would discuss the effects of coal on antelope rangelands or how to protect the endangered prairie grouse from intercontinental missiles stored beneath the buffalo grass at an air force base outside of town. From their home typewriters, they organized letter-writing campaigns aimed at state and federal legislators and copied petitions and took them to church basements to convince fellow parishioners to help preserve the natural world around them.

Preservationists in the 1970s could be found in many American communities. Founded on the vision of leaders like Rachel Carson, they galvanized citizens to save the "big country" to hunt, fish, and recreate in but also because the beauty of open space was so captivating in the way it dwarfed the human in it. We were part of the cosmos when we climbed Rocky Mountain peaks, traveled the interstate across the vast Red Desert, or stood at the mouth of the Mississippi, Arkansas, or Porcupine Rivers. These moments made us realize the scale of nature, and we welcomed the feeling that it was far bigger than us; we were in awe, and our sense of wonder was inflamed.

Rachel Carson was also awed by the nature around her. She drew inspiration from the annual sight of migrating hawks forming kettles above her childhood home and, later, from the tidal pools along the Atlantic Ocean that served as literary muses. In addition to awe, Carson helped us see that industrialization had consequences more dangerous than the visible effects Marsh wrote about. These were quieter effects that stole the spring when pollinators died or muted the crickets in the fall. We now realize that the chemicals Carson warned us about can travel globally, affecting preserved and nonpreserved landscapes far distant from where the emissions were first released.

One of those pollutants is carbon dioxide, a gas whose concentration was first being measured on Mauna Loa, Hawaii, while Carson testified in Congress against the use of pesticides. Few of the atmospheric scientists who tracked carbon dioxide increases predicted the rapid climatological change the world has experienced and none, to my knowledge, predicted that those changes, from levels of 0.028 percent to 0.040 percent today, would drive evolution.

We know that changing climatic conditions, whether the

change is anthropogenic or not, can act as selective agents on populations. For instance, the researchers Rosemary and Peter Grant famously showed that drought in the Galapagos Islands can drive genetically based morphology in birds that survive rapidly changing conditions, ones brought on by the long-standing patterns of El Niño and La Niña. As students of evolution, the Grants illustrated that the beaks on Darwin's finches change as drought mediates the availability and accessibility of seeds; small-billed finches specialize on smaller seeds while large-billed finches are morphologically suited to optimize larger ones. The Grant's long-term data is powerful because it shows that evolution proceeds in the here and now, not only when tectonic plates collide and break away. Furthermore, we know that drought, a signature effect of anthropogenic climate change and a condition that will become more frequent in the Age of Humans, can be a force of evolution.

*   *   *

Not far from where I live is a state park. Lakes, acid bogs, and wetlands dominate the park's landscape, one that was created during the height of the Civilian Conservation Corps when unemployed men built the infrastructure for our national parks as part of the New Deal America. Here, they carefully crafted stone fireplaces and lean-tos to accommodate visitors, testaments to the corps' investment in preserving the natural landscape for humans to enjoy.

Along the edge of one of the lakes is a population of pitcher plants. At the base of each is a cylinder of fused leaves that fill with water and create an aquatic microcosm with its own ecological world. In this world, live pitcher plant mosquitoes. Females lay eggs, larvae hatch, and adults mature and mate again. And when the days shorten and the sunlight dims on the northern latitudes where pitcher plants and their mosquitoes coexist, the plants

Figure 20. Pitcher plants and the mosquitoes that inhabit them exhibit biological evolution in response to changing seasonality brought on by anthropogenic climate change. In the past, cosmic rhythms were the evolutionary pressure driving diapause in mosquitoes. Now these insects are responding to human-induced changes in seasonality. Photo credit: Amy Seidl.

die back and the insects enter their winter diapause until Earth's annual rhythm of warmth and sun returns.

Pitcher plant ecologist William Bradshaw has studied this system for half a century. His expertise is not only the coevolution of plants and insects but also the biological effect of the two great cosmic rhythms that organisms detect: the daily rhythm of Earth rotating on its axis and the annual rhythm of Earth's revolution around the sun. What Bradshaw found is that mosquitoes track seasonal changes, such as the arrival of winter, not via temperature, but through day length, a reliable measurement of the infinitely coupled rhythm of Earth and sun. As a consequence of climate change, however, autumn at northern latitudes is longer than it once was and consequently winter comes later too. So it may come as no surprise that mosquitoes living in pitcher plants have adapted by changing their behavior. And, like the finch's beak, the response is evolutionary.

To take advantage of longer, warmer autumns, pitcher plant mosquitoes have shifted when they enter diapause. Now, rather than diapausing at 15 hours of day length, a trigger that once corresponded to the threat of lethal, winter temperatures, selection favors individuals who diapause at 14 hours, a behavior that increases individual fitness. By sequencing the mosquito's genome and understanding the regions that code for the insect's biological clock, Bradshaw shows that natural selection is acting on the gene that regulates the seasonal response, that is, the earth's revolution around the sun. In essence, mosquitoes are now biologically tracking not the cosmos but humans.

While pitcher plant mosquitoes are one of the first examples of anthropogenic climate change driving biological evolution, they are not the first example of human's evolutionary effect on nature. For instance, with 50 percent of the world's population living in urban areas, it is safe to assume that life inhabiting landscapes alongside ours experiences the exertion of human pressure as well. Nature, after all, lives in urban environments too, and the more we understand the genomics of our coinhabitants, the more we find evidence that they are adapting to our presence.

Humans, as it turns out, have inadvertently begun a great natural experiment. Just as Darwin postulated when he defined adaptation as the response to changing conditions, species respond through selection to new environments. For instance, ants, perhaps the exemplar of an urban adapter, thrive in the urban matrix of lawn and cement, the world's fastest growing biome, and as such have evolved a tolerance for climatic extremes. This predisposes them to move with human populations and track the warmer and more variable temperatures that will come with climate change. Similarly, the carp-like killifish evolve quickly to dioxin-like pollutants including polychlorinated biphenayls, or PCBs, a class of persistent organics whose production was banned in 1979. An evolution for the tolerance of these otherwise lethal chemicals allows killifish to expand into aquatic habitat polluted by humans, a paradox that Rachel Carson would have found both disturbing and biologically intriguing.

Like the evolution of biota to changing environments, stewardship in the Age of Humans must undergo its own evolution given the degree of effect we are having. In earlier times, preservationists were moved to maintain and restore ecological function, to let rivers flow unimpeded and to free prairies from intercontinental missiles. Now there is no landscape free from the perturbation of humans and so we grapple with the wholesale reality of our presence on Earth as signified, in part, by an evolutionary effect that rivals cosmic rhythms.

No condition ever remains the same. This long-standing ecological maxim will define us as we chart the future of nature and our place in it. Ironically, I suggest that that future will lead to a greater sense of wonder, faith even, in the limitlessness of nature and the never-ending evolution of life; the very quality that served as Rachel Carson's muse will become the cornerstone of stewardship in the Age of Humans.

Coda

Bill McKibben

The idea that we now lived on a human-dominated planet first struck me in 1988, after listening to James Hansen's testimony to Congress that is usually regarded as the onset of public discussion of climate change. The conceit that our footprint was now evident everywhere undergirded my 1989 book *The End of Nature*. To wit:

> An idea, a relationship, can go extinct, just like an animal or a plant. The idea in this case is "nature," the separate and wild province, the world apart from man to which he adapted, under whose rules he was born and died. In the past, we spoiled and polluted parts of that nature, inflicted environmental "damage." But that was like stabbing a man with toothpicks: though it hurt, annoyed, degraded, it did not touch vital organs, block the path of lymph or blood. We never thought that we had wrecked nature. Deep down, we never really thought we could: it was too big and too old; its forces—the wind, the rain, the sun—were too strong, too elemental. But, quite by accident, it turned out that the carbon dioxide and other gases we were producing in our pursuit of a better life . . . could alter the power of the sun, could increase its heat. And that increase could change the patterns of moisture and dryness, breed storms in new places, breed deserts . . . We have produced the carbon dioxide—we are ending nature. (41)

The greenhouse effect is a more apt name than those who coined it imagined. The carbon dioxide and trace gases act like the panes of glass on a greenhouse — the analogy is accurate. But it's more than that. We have built a greenhouse, "a human creation" where once there bloomed a sweet and wild garden. (78)

I was glad when, a decade later, others wiser than me gave this understanding a name, the Anthropocene. It struck me then, as now, as a concise and useful shorthand for expressing an idea I've never heard anyone successfully counter: that the world now expresses in every aspect save perhaps the tectonic and volcanic the mark of our heavy-handed species. In fact, this seems so obvious as to be by now a commonplace: we have grown so large that we now cast a planetary shadow. It's the fact of our time. The main question is how to react, and the basic possibilities seem binary to me: become larger still, or try to shrink.

Growing larger is probably the default position, the one our culture and economy naturally seeks out. We're used to the idea of centralization and control, and so things like fracked natural gas and nuclear power plants and patented seeds fit naturally into power structures. But they also fit easily into psychological structures: the best argument for them is, "we're not going to change fast enough for anything else to make a difference."

Which is a good argument if the alternative is to shrink in the approved Thoreauvian manner. Short of crisis, it's hard for me to imagine that we'll embrace an ascetic life in numbers substantial enough to matter (though I'm convinced by the argument that we'd be happier in many ways if we did). Too many people are coming into the sudden possibility of abundance, and its short-term allure is too strong, for austerity to be a persuasive choice.

Happily, it seems to me there's another axis we might work on, one different from the big/small dichotomy. For my money the great writer and thinker of our time is Wendell Berry, the Kentucky farmer and essayist who has argued persuasively

that the direction we want to move in is toward community—
community with the soil, community with the rest of creation,
community with the people who live around us. His moral and
practical argument fits nicely with an array of new insights about
human psychology and physiology: we're happier and healthier
when we have strong social connections.

It's those connections, of course, that have atrophied in the
high consumer society that we've built. As we've become bigger *as individuals* our social networks have steadily shrunk: the
average American, for instance, has half as many close friends as
a half century ago, which probably helps explain why our growing affluence hasn't produced transports of ecstasy in our civilization. It also explains the sheer size of our carbon footprint: our
hyperindividualism is evident in the chemistry of every cubic
meter of air.

Is it possible that there's a way to build a different world not
by putting up nuclear plants but by building up networks? We
have, obviously, a new tool in the Internet, the one new wild card
of the last decades. In its operation, of course, it can be as iso-
lating as empowering (and that's why, say, Berry doesn't care for
it). But it also operates as a metaphor, again the great new metaphor of our time. The *network* is perhaps the goal for a planet that
has scant options.

In practical terms, consider energy. We could build some
nuclear power plants, and they might or might not come in time
to do something about climate change. (Given their endless
complexity, cost overruns, and delays one tends to doubt it, but
perhaps.) Or you could build out a huge network of solar panels
and windmills. It's doable—a 2013 study from the University of
Delaware found that by 2030 such a robust network could keep
American lights on 99.9 percent of the time. It won't be free,
but costs are falling quickly, so that study and others have found
the cost won't necessarily be ruinous. Storage is a problem, but
probably not an insurmountable one: new innovations come online daily. In fact, the almost incomprehensibly rapid growth of

the Internet seems to offer real hope that energy systems too could metastasize quickly enough to actually matter against the timetable set by the physics of global warming

And the upside of such a network (beyond the fall in carbon emissions) is precisely that it is a network—that it enmeshes us in a community of sorts. I have solar panels all over my roof, tied to our grid; I like the idea that my neighbor cools his beer before the Red Sox game with the sunlight falling on my shingles. This is the same kind of affection that people across the country (and not just in wealthy or crunchy enclaves) have found for local food and the farmer's market. Connection is a benefit in and of itself, and building an economy that emphasizes connection instead of isolation seems like a practical task that might produce a virtuous cycle.

I have more faith in the idea of large-scale cooperation than I do in elite management—more faith in the Internet and the beehive than in Monsanto and the nuclear industry. We are on the edge of a climate apocalypse, and it's crucial we make quick choices in directions that have at least some chance of prevailing. I could be wrong—there's a good deal to be said for working with the momentum of our existing, highly centralized economy—but I like the math of multiplying networks.

Another way of saying this is: I bet that the great intellectual insight of the twentieth century won't turn out to be nuclear physics. I bet it will turn out to be ecology, the idea that all things are deeply interconnected. In the casino that is our perilous age I'd stack my chips on that bet, doing everything possible to make it pay off (a price on carbon, a huge subsidy for renewables, a pitched fight against the corporate powers of centralization). I'm not confident the bet will pay off—the cheerful title of that 1989 book, after all, was *The End of Nature*. But I'm not hopeless either. We've got a wild card, and it's time to play it.

Notes

Minteer and Pyne, Writing on Stone, Writing in the Wind

The American preservationist tradition is explored in Rod Nash's *Wilderness and the American Mind*, 5th ed. (New Haven: Yale University Press, 2014), Stephen Fox's *The American Conservation Movement: John Muir and His Legacy* (Madison: University of Wisconsin Press, 1986), and Don Worster's *Nature's Economy: A History of Ecological Ideas*, 2nd ed. (Cambridge: Cambridge University Press, 1994). John McPhee's *Encounters with the Archdruid* (New York: Farrar, Straus and Giroux, 1980) provides the indelible account of David Brower's efforts to preserve the Grand Canyon from development. For discussions of the scientific, ethical, and management challenges of setting targets for conservation, restoration, and preservation under conditions of rapid ecological change, see John W. Williams and Stephen J. Jackson, "Novel Climates, No-Analog Plant Communities, and Ecological Surprises: Past and Future," *Frontiers in Ecology and Evolution* 5 (2007): 475–82; Stephen T. Jackson and Richard J. Hobbs, "Ecological Restoration in the Light of Ecological History," *Science* 325 (2009): 567–69; Alejandro Camacho, Holly Doremus, Jason S. McLachlan, and Ben A. Minteer, "Reassessing Conservation Goals in a Changing Climate," *Issues in Science and Technology* 26 (2010): 21–26; and Ben A. Minteer and James P. Collins, "Species Conservation, Rapid Environmental Change, and Ecological Ethics," *Nature Education Knowledge* 3 (2012): 4. For a scientific survey of the human modification of earth systems, see Erle C. Ellis, "Anthropogenic Transformation of the Terrestrial Biosphere," *Philosophical Transactions of the Royal Society* A 369 (2011): 1010–35.

Paul J. Crutzen's early paper on the Anthropocene construct,

"Geology of Mankind," appeared in *Nature* 415 (2002): 23. See also W. Steffen, J. Grinevald, P. Crutzen, and J. McNeill, "The Anthropocene: Conceptual and Historical Perspectives," *Philosophical Transactions of the Royal Society* A 369 (2011): 842–67. Optimistic assessments of the Anthropocene idea and its implications for rethinking nature conservation include Peter Kareiva, Michelle Marvier, and Robert Lalasz, "Conservation in the Anthropocene: Beyond Solitude and Fragility," *Breakthrough Journal* (http://thebreakthrough.org/index.php /journal/past-issues/issue-2/conservation-in-the-anthropocene/). A more skeptical view may be found in the similarly titled article by Tim Caro et al., "Conservation in the Anthropocene," *Conservation Biology* 26 (2011): 185–88; and in Michael Soulé, "The New Conservation," *Conservation Biology* 27 (2013): 895–97. See also the recent collection (edited by George Wuerthner et al.), *Keeping the Wild: Against the Domestication of the Earth* (Washington, DC: Island Press, 2014), which offers a robust critique of the Anthropocene idea from the vantage point of traditional nature preservationism.

Leopold's line about writing on the land appears in his essay "Axe-in-Hand" in *A Sand County Almanac* (Oxford: Oxford University Press, 1949), 68. The quote from John McPhee's *Coming into the Country* (New York: Farrar, Straus and Giroux 1991) appears on p. 426.

Revkin, Restoring the Nature of America

Robert F. Kennedy's speech on gross domestic product may be found at http://www.jfklibrary.org/Research/Research-Aids/Ready-Reference /RFK-Speeches/Remarks-of-Robert-F-Kennedy-at-the-University -of-Kansas-March-18-1968.aspx. Richard M. Nixon's 1970 State of the Union address is accessible at http://www.presidency.ucsb.edu/ws /?pid=2921.

McNeill, Nature Preservation and Political Power in the Anthropocene

New Anthropocene journals include *Journal of the Anthropocene*; *Anthropocene Review*; *Elementa: Journal of Anthropocene Science*. For Buffon's remarks, see M. le Comte de Buffon, *Les Epoques de la Nature*,

Supplément, Tome V, *Histoire naturelle, générale et particulière* (Paris: Imprimerie royale, 1778). Stoppani's views appear in *Corso di geologia* (1873). The sustained use of the term "Anthropocene" dates from P. J. Crutzen and E. G. Stoermer, "The Anthropocene," *IGBP Newsletter* 41 (2000): 12.

Rival views of the onset of the Anthropocene appear in William Ruddiman, *Earth Transformed* (New York: W. H. Freeman, 2014), 231–43; John Gowdy and Lisi Krall, "The Ultrasocial Origin of the Anthropocene," *Ecological Economics* 95 (2013): 137–47; G. Certini and R. Scalenghe, "Anthropogenic Soils Are the Golden Spikes for the Anthropocene," *The Holocene* 21 (2011): 1269–74; W. Steffen, J. Grinevald, P. Crutzen, and J. McNeill, "The Anthropocene: Conceptual and Historical Perspectives," *Philosophical Transactions of the Royal Society* A 369 (2011): 842–67; E. Ellis, K. Goldewijk, S. Klein Siebert, D. Lightman, and N. Ramankutty, "Anthropogenic Transformation of the Biomes 1700 to 2000," *Global Ecology and Biogeography* 19 (2011): 589–606; Jan Zalasiewicsz, "The Epoch of Humans," *Nature GeoScience* 6, 8–9 (2013), doi:10.1038/ngeo1674; Christian Pfister, "Das 1950er Syndrom: Die Epochenschwelle der Mensch-Umwelt-Beziehung zwischen Industriegesellschaft und Konsumgesellschaft," *GAIA — Ecological Perspectives for Science and Society*, 3, no. 2 (1994): 71–90.

For a sampling of the stratigraphers' debates, see W. Autin and J. Holbrook, "Is the Anthropocene an Issue of Stratigraphy or Pop Culture?," *GSA Today* 22 (2012): 60–61; John Lewin and Mark G. Macklin, "Marking Time in Geomorphology: Should We Try to Formalise an Anthropocene Definition?," *Earth Surface Processes and Landforms*, October 17, 2013, doi:10.1002/esp.3484; J. Zalasiewicz et al, "Stratigraphy of the Anthropocene," *Philosophical Transactions of the Royal Society* A 369 (2011): 1036–55; and Antony G. Brown et al., "The Anthropocene: Is There a Geomorphological Case?," *Earth Surface Processes and Landforms* 38(2013): 431–34.

For moral and political critiques of the concept of the Anthropocene, see Agnès Sinaï, ed., *Penser la décroissance. Politiques de l'Anthropocène* (Paris: Les Presses de Sciences Po, 2013); Andreas Malm and Alf Hornborg, "The Geology of Mankind? A Critique of the Anthropocene Narrative," *Anthropocene Review* 1 (2014): 62–69; Tim Caro et al., "Conservation in the Anthropocene," *Conservation Biology* 26 (2011): 185–88;

Tim Caro, "Anthropocene: Keep the Guard Up," *Nature* 502 (2013), doi:10.1038/502624a.

For an interesting analysis of the tendency of power alternately to centralize and decentralize, see Adam Watson, *The Evolution of International Society: A Comparative Historical Analysis* (London: Routledge, 1992).

Rolston, After Preservation? Dynamic Nature in the Anthropocene

Gifford Pinchot's "The Fight for Conservation" is reprinted in Donald Worster, ed., *American Environmentalism: The Formative Period, 1860–1915*, 84–95 (New York: John Wiley, 1973). John Muir's praise of wilderness cathedrals is in *The Yosemite* (Garden City, NY: Doubleday and Company, [1912] 1965). Baird Callicott revisits wilderness in "The Wilderness Idea Revisited," *Environmental Professional* 13 (1991): 235–47. Aldo Leopold's "Land Ethic" is in his *A Sand County Almanac* (New York: Oxford University Press, [1949] 1968). Erle Ellis sees "The Planet of No Return" in *Breakthrough Journal* 2 (Fall 2011): 39–44. He joins colleagues in a forward-looking environmentalism in the *New York Times*, December 8, 2011, A39. The American Geosciences book is George A. Seielstad, *Dawn of the Anthropocene: Humanity's Defining Moment* (Alexandria, VA: American Geosciences Institute, 2012, a digital book).

Richard Alley provides us with *Earth: The Operator's Manual* (New York: W. W. Norton, 2011). Mark Lynas celebrates our becoming "the God species" in *The God Species: Saving the Planet in the Age of Humans* (Washington, DC: National Geographic, 2011). Allen Thompson has "Radical Hope for Living Well in a Warmer World" in the *Journal of Agricultural and Environmental Ethics* 23 (2010): 43–59. Brad Allenby finds the biosphere increasingly a human product in IEEE Technology and Society Magazine, Winter 2000/2001. Richard Hobbs and colleagues address human-made ecosystems in *Novel Ecosystems: Intervening in the New Ecological World Order* (Oxford: Wiley Blackwell, 2013). The *Economist* theme issue is "Welcome to the Anthropocene," May 28, 2011, vol. 399, no. 8735.

Johan Rockström analyzes "A Safe Operating Space for Humanity" in *Nature* 461 (September, 24, 2009): 472–75. George Peterken assesses the multiple levels of naturalness in *Natural Woodland: Ecology and Conservation in Northern Temperate Regions* (Cambridge: Cambridge University Press, 1996). Gregory Aplet's zones of wildness are found in "On the Nature of Wildness: Exploring What Wilderness Really Protects," *University of Denver Law Review* 76 (1999): 347–67. J. M. McCloskey and Heather Spalding document global wilderness in "A Reconnaissance Level Inventory of the Amount of Wilderness Remaining in the World," *Ambio* 18 (1989): 221–27. Peter Kareiva and Michelle Marvier scrap conserving biodiversity unless it is good for people in "Conservation for the People," *Scientific American* 297 (October 2007): 50–57. The Royal Society advocates more intensely exploiting nature in *Reaping the Benefits: Science and the Sustainable Intensification of Global Agriculture* (London: Royal Society, 2009).

Marris, Humility in the Anthropocene

On the discussion of extinction, philosophers (and others) may wonder whether I am claiming that *every* extinction is morally bad. What about smallpox? What if you had to decide between an extinction and killing a bunch of humans? These are good questions, and there probably are some very limited exceptions to the basic rule that it is wrong to let extinctions happen. But generally speaking, where the choice is to stop extinction or to let it happen, with no other complicated factors, we can all agree one ought to stop it. For the essay by Michael Soulé, see http://www.michaelsoule.com/frontpage_files/73/73_frontpage_file2.doc.

For Postell, see http://newswatch.nationalgeographic.com/2013/06/03/water-and-us-in-the-anthropocene/. The Kingsnorth essay may be found at http://www.orionmagazine.org/index.php/articles/article/7277. Wapner's essay is available at http://www.tikkun.org/nextgen/humility-in-a-climate-age.

Foreman, The Anthropocene and Ozymandias

My extinction discussion is in Dave Foreman, *Rewilding North America* (Washington, DC: Island Press, 2004). Leopold's comment on the "degree of wildness" is in Aldo Leopold, *A Sand County Almanac* (New

York: Oxford University Press, 1949), 189. Senator Church's statement is in Frank Church, "The Wilderness Act Applies to the East," *Congressional Record—Senate*, January 16, 1973, 737. I've taken the four definitions of wilderness directly from the 1964 Wilderness Act. I've stolen the "hand of Man" quote from the late Dave Brower. "Overkill" comes from the late paleontologist Paul Martin and "Big Hairies" from the paleontologist Peter Ward. The quotation from Peter Kareiva and coauthors about the passenger pigeon's extinction is found in Peter Kareiva, Michelle Marvier, and Robert Lalasz, "Conservation in the Anthropocene," *Breakthrough Journal*, Fall 2011, 29–37. The quote from Gould is in Stephen Jay Gould, "Reconstructing (and Deconstructing) the Past," in Stephen Jay Gould, ed., *The Book of Life* (New York: W. W. Norton, 2001), 10. Leopold's "wild things" comment is in Leopold, *A Sand County Almanac*, vii. *Ozymandias* is from *The Complete Poems of Percy Bysshe Shelley* (New York: Modern Library, 1994), 589.

Worster, The Higher Altruism

Charles Darwin puzzled over why individuals might sacrifice themselves for the welfare of another, contrary to the law of competition in natural selection, an anomaly he feared that might be fatal to his theory of evolution (*On the Origin of Species* [1859], 236). In recent decades biologists have tried to answer this challenge, but a debate is now raging between Richard Dawkins and Edward O. Wilson over how altruism might emerge from natural selection. Dawkins argues that altruism can only evolve out of self-interest—see his *The Selfish Gene* (1976)—while Wilson's book *The Social Conquest of the Earth* (2012) takes the more radical view that selection can work at the group as well as individual level, particularly among "social species" like ants and humans. I am drawn more to Wilson's ideas, but regardless of who is right the altruism I have in mind requires a conscious intent to help others, and in this case to help protect the otherness of the nonhuman human world—and such intention is the result of cultural, not simply biological, evolution.

But a key question for preservation to ask is where does nature's otherness end? Does man-made climate change destroy all that autonomy and make preservation meaningless? Perhaps a better test than unintended and unpredictable climate change is how much of the biosphere's production goes to support human life. One useful calculation comes from Peter Vitousek, Paul Ehrlich, Anne Ehrlich, and Pamela Matson, "Human Appropriation of the Products of Photosynthesis," *BioScience* 36 (1986): 368–73. Out of 224 petagrams of biomass produced on the planet each year through photosynthesis, they calculate that humans and their livestock directly consume only seven petagrams. If one adds a wider "impact" from cities, croplands, pastures, and tree plantations, then humans could be appropriating as much as 40 percent of all terrestrial net productivity — or much less, depending on the method of calculation. Those figures would seem to leave a lot of room for the work of preserving untrammeled nature.

Preservationists may have more, not fewer, opportunities in the future. The UN Department of Economic and Social Affairs, "World Population to 2300" (New York, 2004), projects a human population peak of 9.22 billion in 2072 and then a slight decline to a long-term steady state for another century or more. But Wolfgang Lutz and Sergei Scherbov of the International Institute for Applied Systems Analysis argue, in "Exploratory Extensions of IIASA's World Population Projections: Scenarios to 2300" (Vienna, 2008), that a massive human population decline may lie in the future, not a steady state. Europe's fertility rate has already fallen to 1.5 children per woman; if the world rate falls to the same level and stays there, then by 2200 the world population will fall to half of what it is today, and by 2300 it will be no more than one billion. If that happens, then much of the earth's surface will no longer be required for agriculture or industry, allowing a vast expansion of wilderness.

Vucetich et al., The Anthropocene: Disturbing Name, Limited Insight

The quote from Kathleen Dean Moore appears in *Earth Island Journal* 28 (2013): 19–20.

For an introduction to the is/ought problem in the context of environmental ethics, see J. Baird Callicott, *Environmental Ethics* 4 (1982): 163–74. Peter Kareiva and Michelle Marvier's discussion of conservation in the Anthropocene appears in *BioScience* 62 (2012): 962–69.

Michael Soulé's early vision for conservation biology appeared in *BioScience* 35 (1985): 727–34. For the arguments of Steffen et al. on the Anthropocene, see "The Anthropocene: Are Humans Now Over-whelming the Great Forces of Nature?," *Ambio* 36 (2007): 614–21. The quote from Steffen et al. ("The Anthropocene is a reminder that the Holocene...") appears in "The Anthropocene: From Global Change to Planetary Stewardship," *Ambio* 40 (2011): 739–61. See "Sustainability: Virtuous or Vulgar?" by John A. Vucetich and Michael P. Nelson, in *BioScience* 60 (2010): 539–44, for a discussion of the multiple interpretations of sustainability. For considerations of the virtues of precaution, humility, empathy, and rationality, see (respectively) Dan F. Doak et al., "Understanding and Predicting Ecological Dynamics: Are Major Surprises Inevitable?," *Ecology* 89 (2008): 952–61; John A. Vucetich and Michael P. Nelson, "The Infirm Ethical Foundations of Conservation," in Marc Bekoff, ed., *Ignoring Nature No More: The Case for Compassionate Conservation* (Chicago: University of Chicago Press, 2013), 9–26; Michael P. Nelson and John A. Vucetich, "Environmental Ethics for Wildlife Management," in D. J. Decker, S. J. Riley, and W. F. Siemer, eds., *Human Dimensions of Wildlife Management* (Baltimore: Johns Hopkins University Press, 2012), 223–37; Michael P. Nelson and John A. Vucetich, "On Advocacy by Environmental Scientists: What, Whether, Why, and How," *Conservation Biology* 23 (2009): 1090–1101; John A. Vucetich and Michael P. Nelson, "What Are 60 Warblers Worth? Killing in the Name of Conservation," *Oikos* 116 (2007): 1267–78; and Amartya Sen, *The Idea of Justice* (Cambridge, MA: Harvard University Press, 2009). On ecosystem health and the requirement of intervention, see John A. Vucetich, Michael P. Nelson, and Rolf O. Peterson, "Should Isle Royale Wolves Be Reintroduced? A Case Study on Wilderness Management in a Changing World," *George Wright Forum* 29 (2012): 126–47.

Norton, Ecology and the Human Future

The editorial by Wei-Ning Xiang in which he advances his notion of "ecological wisdom" appears in *Landscape and Urban Planning* 121 (2014): 65–69. Rittel and Webber's original formulation of "wicked problems" may be found in their paper "Dilemmas in a General Theory of Planning," *Policy Sciences* (1973): 155–69.

Meine, A Letter to the Editors: In Defense of the Relative Wild

Despite my qualms, I really do thank the editors, Ben Minteer and Stephen Pyne, for the inviting me to contribute to this volume. I also thank Dr. Paul Zedler of the University of Wisconsin–Madison for bringing to my attention T. C. Chamberlin's 1883 statement in the *Geology of Wisconsin*. Thanks as well to George Archibald for allowing me to use his photo of Gangkhar Puensum.

The "prophetic observers and philosophers of Earth history" are reviewed in Will Steffen, Jacques Grinevald, Paul Crutzen, and John McNeill, "The Anthropocene: Conceptual and Historical Perspectives," *Philosophical Transactions of the Royal Society* A 369 (2011): 842–67. T. C. Chamberlin's discussion of the "Psychozoic Era" can be found in *Volume I* of *Geology of Wisconsin, Survey of 1873–1879*, Commissioners of Public Printing (Wisconsin, 1883), 299–300. See also James R. Fleming, "T. C. Chamberlin, Climate Change, and Cosmogony," *Studies in History and Philosophy of Science Part B: Studies in History and Philosophy of Modern Physics* 31 (2000): 293–308. For the "myth of the pristine," see William M. Denevan's original "The Pristine Myth: The Landscape of the Americas in 1492," *Annals of the Association of American Geographers* 82 (1992): 369–85; and his subsequent review, "The Pristine Myth Revisited," *Geographical Review* 101 (2011): 576–91. Thomas Vale suggests the need to examine the "myth of the humanized" in the introduction to his book *Fire, Native Peoples, and the Natural Landscape* (Washington, DC: Island Press, 2002).

The text of the 1909 *Report of the Commission on Country Life* is available online at several websites. See also William L. Bowers, "Country-Life Reform, 1900–1920: A Neglected Aspect of Progressive Era History," *Agricultural History* 45, no. 3 (1971): 211–21; and Scott J. Peters and Paul A. Morgan, "The Country Life Commission: Reconsidering a Milestone in American Agricultural History," *Agricultural History* 78, no. 3 (2004): 289–316. Bennett's quotation is from the landmark publication (coauthored with William Ridgely Chapline), *Soil Erosion: A National Menace*, US Department of Agriculture Circular No. 33 (Washington, DC: US Government Printing Office, 1928). Leopold's phrase is from his introduction to *A Sand County Almanac and Sketches Here and There* (New York: Oxford University Press, 1949), ix. Mumford of course wrote extensively on the development of cities and the quality of urban life; the quotation is from *The Culture of Cities* (New York: Harcourt, Brace, 1938). Ben Minteer explores this alternative channel of conservation history in depth in *The Landscape of Reform: Civic Pragmatism and Environmental Thought in America* (Cambridge, MA: MIT Press, 2006). See also Paul B. Thompson, "Expanding the Conservation Tradition: The Agrarian Vision," chap. 6 in Ben A. Minteer and Robert E. Manning, eds., *Reconstructing Conservation: Finding Common Ground* (Washington, DC: Island Press, 2003).

Leopold's statement on wilderness as a "relative condition" occurs in "Wilderness as a Form of Land Use," *Journal of Land and Public Utility Economics* 1 (1925): 398–404. Leopold expressed the same idea in a number of other manuscripts and publications. His reference to the "weeds in a city lot" is from "Conservation Esthetic," *Bird Lore* 40 (1938): 101–9, and subsequently included in *A Sand County Almanac*, 174. For my own take on conservation "across the landscape," see "Leopold's Fine Line," chap. 4 in *Correction Lines: Essays on Land, Leopold, and Conservation* (Washington, DC: Island Press, 2004); and "Crossing the Great Divide," *Quivira Coalition Journal* 30 (2007): 3–11. Wendell Berry's quotation is from "Out of Your Car, Off Your Horse," *Atlantic* 267 (1991): 61–63. Among the many recent volumes that look "beyond preservation," see Richard L. Knight and Peter Landres, eds., *Stewardship across Boundaries* (Washington, DC: Island Press, 1998); Nathan F. Sayre, *Working Wilderness: The Malpai Borderlands Group and the Future of the Western Range* (Tucson, AZ: Rio Nuevo Publishers, 2005); and Richard L. Knight and Courtney White, eds., *Conservation for a New Generation* (Washington, DC: Island Press, 2008).

George Archibald sent his e-mail message from Bhutan on November 18, 2013. The late Peter Matthiessen wrote of Bhutan's black-necked cranes in his book *The Birds of Heaven: Travels with Cranes* (New York: Macmillan, 2001). A recent review of the species' status there is P. H.

U. R. B. A Lhendup and Edward L. Webb, "Black-Necked Cranes *Grus nigricollis* in Bhutan: Migration Routes, Threats and Conservation Prospects," *Forktail* 25 (2009): 125–29. For a relevant commentary on Gangkhar Puensum, see Martin Lin, "Preserving the Last Shangri-la: Responsible Travels through Bhutan," http://archive.is/w84P (accessed December 4, 2013).

Minteer, When Extinction Is a Virtue

Leopold's discussion of the passenger pigeon appears in his essay "On a Monument to the Pigeon," in *A Sand County Almanac* (New York: Oxford University Press, 1949). His remark about the "oldest task in human history" is from his 1938 essay "Engineering and Conservation," collected in S. L. Flader and J. B. Callicott, eds. *The River of the Mother of Good and Other Essays by Aldo Leopold* (Madison: University of Wisconsin Press), quote on 254. On the potential pros and cons of the de-extinction idea, see J. S. Shankow and H. T. Greely, "What if Extinction Is Not Forever?," *Science* 340 (2013): 32–33; Carl Zimmer, "Bringing Them Back to Life," *National Geographic*, April 2013; Ronald Sandler, "The Ethics of Reviving Long Extinct Species," *Conservation Biology* 28 (2013): 354–60; and Nathaniel Rich, "The Mammoth Cometh," *New York Times Magazine*, February 27, 2014. Stewart Brand's version of ecopragmatism is put forward in his *Whole Earth Discipline: An Ecopragmatist Manifesto* (New York: Viking, 2009). For Brand's brief on de-extinction, see his Tedx presentation, "The Dawn of De-extinction: Are You Ready?," http://www.ted.com/talks/stewart_brand_the_dawn _of_de_extinction_are_you_ready.html, as well as his interesting debate with Paul Ehrlich in the online magazine *Yale Environment 360*, http://e360.yale.edu/feature/the_case_for_de-extinction_why_we_should _bring_back_the_woolly_mammoth/2721/.

The quote from Shellenberger and Nordhaus appears in their 2009 book *Break Through: Why We Can't Leave Saving the Environment to the Environmentalists* (New York: Mariner Books, 2009), 272. John Dewey's discussion of "natural piety" may be found in his 1934 book *A Common Faith*, collected in volume 9 of J. A. Boydston, ed., *The Later Works of John Dewey, 1925–1953* (Carbondale: Southern Illinois University Press,

1986). The remark from *The Public and Its Problems*, which is collected in volume 2 of Boydston, ed., *The Later Works of John Dewey, 1925–1953* (Carbondale: Southern Illinois University Press, 1984), appears on 345.

Thanks to Curt Meine for sharing his thoughts on an earlier version of this essay and for bringing the Leopold quote from *Game Management* to my attention.

An alternate, much shorter version of this essay appeared in *Nature* ("Is It Right to Reverse Extinction?," *Nature* 509 [2014]: 261).

Greene, Pleistocene Rewilding and the Future of Biodiversity

Core publications referred to are C. J. Donlan, H. W. Greene, J. Berger, C. E. Bock, J. H. Bock, D. A. Burney, J. A. Estes, D. Foreman, P. S. Martin, G. W. Roemer, F. A. Smith, and M. E. Soulé, "Re-wilding North America," *Nature* 436 (2005): 913–14; C. J. Donlan, J. Berger, C. E. Bock, J. H. Bock, D. A. Burney, J. A. Estes, D. Foreman, P. S. Martin, G. W. Roemer, F. A. Smith, M. E. Soulé, and H. W. Greene, "Pleistocene Rewilding: An Optimistic Agenda for Twenty-First-Century Conservation," *American Naturalist* 168 (2006): 660–81; and C. J. Donlan and H. W. Greene, "NLIMBY: No Lions in My Backyard," in M. Hall, ed., *Restoration and History: The Search for a Usable Environmental Past* (Routledge, 2010), 293–305. The paraphrased passage and quote are from Donlan et al., "Pleistocene Rewilding," 674.

Fiege, The Democratic Promise of Nature Preservation

On Turner, see John Mack Faragher, ed., *Rereading Frederick Jackson Turner* (New York: Henry Holt, 1994). On political history from the perspective of the Anthropocene, see, for example, Geoffrey Parker, *Global Crisis: War, Climate Change, and Catastrophe in the Seventeenth Century* (New Haven: Yale University Press, 2013). On the need to preserve democracy in the Anthropocene, see Clive Hamilton, *Requiem for a Species: Why We Resist the Truth about Climate Change* (New York: Earthscan, 2010). On evolving ideas of preservation, see, for example, Mark Fiege, *The Republic of Nature: An Environmental History of the United States* (Seattle: University of Washington Press, 2012); Drew McCoy, *The Elusive Republic: Political Economy in Jeffersonian America*

(Chapel Hill: University of North Carolina Press, 1980); Denise D. Meringolo, *Museums, Monuments, and National Parks: Toward a New Genealogy of Public History* (Amherst: University of Massachusetts Press, 2012); Roderick Nash, *Wilderness and the American Mind*, 3rd ed. (New Haven: Yale University Press, 1982); Roderick Nash, *The Rights of Nature: A History of Environmental Ethics* (Madison: University of Wisconsin Press, 1989); and Lary M. Dilsaver, *America's National Park System: The Critical Documents* (Lanham, MD: Rowman and Littlefield, 1994). On the exclusivity of early parks and protected areas, see Karl Jacoby, *Crimes against Nature: Squatters, Poachers, Thieves, and the Hidden History of American Conservation* (Berkeley: University of California Press, 2003); Louis S. Warren, *The Hunter's Game: Poachers and Conservationists in Twentieth Century America* (New Haven: Yale University Press, 1997); and Jen A. Huntley, *The Making of Yosemite: James Mason Hutchings and the Origin of America's Most Popular National Park* (Lawrence: University Press of Kansas, 2011). On Muir, see Donald Worster, *A Passion for Nature: The Life of John Muir* (New York: Oxford University Press, 2008); and John Muir, *The Eight Wilderness Discovery Books*, ed. Terry Gifford (Seattle: The Mountaineers, 1992). On modernist separations, see Charles S. Maier, "Consigning the Twentieth Century to History: Alternative Narratives for the Modern Era," *American Historical Review* 105 (June 2000): 807–31. On the separation of human history from natural history, see Clive Hamilton, *Earthmasters: The Dawn of the Age of Climate Engineering* (New Haven: Yale University Press, 2013). On human virtue and democracy cultivated in service to nonhuman nature, see Holmes Rolston III, *Genes, Genesis, and God: Values and Their Origins in Natural and Human History* (New York: Cambridge University Press, 1999), and Jodi Hilty, William Z. Lidicker Jr., and Adina M. Merenlender, *Corridor Ecology: The Science and Practice of Linking Landscapes for Biodiversity Conservation* (Washington, DC: Island Press, 2006). On philosophical pragmatism, see, for example, Ben A. Minteer, *The Landscape of Reform: Civic Pragmatism and Environmental Thought in America* (Cambridge, MA: MIT Press, 2006).

Marvier and Wong, Move Over Grizzly Adams —
Conservation for the Rest of Us

Data regarding the diversity of scientists in the US Department of Interior are provided by J. S. Burrelli and J. C. Falkenheim "Diversity in the Federal Science and Engineering Workforce" (Arlington, VA: Division of Science Resources Statistics, National Science Foundation, 2011), http://www.nsf.gov/statistics/infbrief/nsf11303. Data regarding the diversity of staff and board members at nongovernmental conservation organizations are provided by D. E. Taylor "The State of Diversity in Environmental Organizations" (Green 2.0, 2014), http://diverse green.org/report. One example of a plea for greater diversity within the fields of conservation and wildlife management can be found at R. Lopez and C. H. Brown, "Why Diversity Matters: Broadening Our Reach Will Sustain Natural Resources," *Wildlife Professional* 5 (2011): 20–27. Examples of public opinion polls finding that people of color and women express greater concern and greater willingness to pay for nature protection include B. Czech, P. K. Devers, and P. R. Krausman, "The Relationship of Gender to Species Conservation Attitudes," *Wildlife Society Bulletin* 29 (2001): 187–94, and D. Metz and L. Weigel, "Key Findings from National Voter Survey on Conservation among Voters of Color," (Fairbank, Maslin, Maullin & Associates and Public Opinion Strategies, 2009), http://www.environmentaldiversity.org/documents/NationalVotersofColorSummary.doc.

We (Michelle Marvier and Hazel Wong) report data from national polls assessing public motivations for conservation in "Winning Back Broad Public Support for Conservation," *Journal of Environmental Studies and Sciences* 2 (2012): 291–95. The decline in people identifying as environmentalists and conservationists is documented by K. Bowman, A. Rugg, and J. Marsico, "Polls on the Environment, Energy, Global Warming, and Nuclear Power," American Enterprise Institute for Public Policy Research, http://www.aei.org/paper/politics-and -public-opinion/polls/polls-on-the-environment-energy-global -warming-and-nuclear-power-april-2013/. Differences in the use of parks and open spaces across demographic groups are document by Paul H. Gobster, "Managing Urban Parks for a Racially and Ethnically Diverse Clientele," *Leisure Sciences* 24, no. 2 (2002): 143–59 and the references therein. The quote from Audrey Peterman and the analysis of images appearing in *Outside* magazine were reported in "What's Right with This Picture? Diversity in the Outdoors," *Outside*, 2011, http://www.outsideonline.com/outdoor-adventure/politics/Whats-Right-with-This-Picture.html.

The assessment of The Nature Conservancy's Leaders in Environmental Action for the Future program was authored by J. Fraser, R. Gupta, and S. J. Rank, "The Nature Conservancy's Leaders in Environmental Action for the Future: 2012 Program Evaluation" (New York: New Knowledge Organization Ltd., 2013). Dave Foreman's appeals to "true conservationists" appear in his book *Take Back Conservation* (Durango, CO: Raven's Eye Press, 2012). Michael Soulé has authored several pieces arguing against a more human-centric framing of conservation, one of which is "The 'New Conservation,'" *Conservation Biology* 27 (2013): 895–97.

Contributors

CHELSEA K. BATAVIA is a PhD student in the Department of Forest Ecosystems and Society in the College of Forestry at Oregon State University in Corvallis.

F. STUART (TERRY) CHAPIN III is professor emeritus of ecology in the Institute of Arctic Biology at the University of Alaska Fairbanks. His most recent book (with Pam Matson and Peter Vitousek) is *Principles of Terrestrial Ecosystem Ecology* (2nd ed.).

NORMAN L. CHRISTENSEN is research professor and founding dean of the Nicholas School at Duke University in Durham, NC. He is author of *The Environment and You*.

JAMIE RAPPAPORT CLARK is president and CEO of Defenders of Wildlife. She is a national expert on the Endangered Species Act and imperiled wildlife and served as director of the US Fish and Wildlife Service from 1997 to 2001.

WILLIAM WALLACE COVINGTON is Regents' Professor in the School of Forestry at Northern Arizona University in Flagstaff. He is director of the Ecological Restoration Institute (ERI), a research, education, and outreach institution at NAU that advises on restoration treatment outcomes, strategies, and techniques in the US Southwest.

ERLE C. ELLIS is professor in the Department of Geography and Environmental Systems at the University of Maryland, Baltimore County. He is an ecologist investigating what scientists increasingly describe as the Anthropocene, the Age of Humans.

MARK FIEGE is professor of history at Colorado State University in

Fort Collins. He is author, most recently, of *The Republic of Nature: An Environmental History of the United States.*

DAVE FOREMAN is cofounder of the Wildlands Project and is executive director of The Rewilding Institute. He is author most recently of *Take Back Conservation.*

HARRY W. GREENE is Stephen Weiss Presidential Fellow and professor of ecology and evolutionary biology at Cornell University in Ithaca, NY. His most recent book is *Tracks and Shadows: Field Biology as Art.*

EMMA MARRIS is a writer based in Klamath Falls, Oregon. She is author of *Rambunctious Garden: Saving Nature in a Post-Wild World.*

MICHELLE MARVIER is professor and chair of the Department of Environmental Studies and Sciences at Santa Clara University in Santa Clara, CA. She is coauthor of *Conservation Science: Balancing the Needs of People and Nature.*

BILL MCKIBBEN is a writer, environmental activist, and scholar in residence at Middlebury College in Middlebury, VT. His most recent book is *Oil and Honey: The Education of an Unlikely Activist.*

J. R. MCNEILL is University Professor in the Department of History at Georgetown University in Washington, DC. He is author of *Something New under the Sun: An Environmental History of the 20th-Century World.*

CURT MEINE is senior fellow at the Center for Humans and Nature in Chicago and at the Aldo Leopold Foundation in Baraboo, WI. He is author of *Aldo Leopold: His Life and Work.*

BEN A. MINTEER is the Arizona Zoological Society Endowed Chair at Arizona State University and professor of environmental ethics and conservation in ASU's School of Life Sciences. He is author most recently of *Refounding Environmental Ethics: Pragmatism, Principle, and Practice.*

MICHAEL PAUL NELSON is the Ruth H. Spaniol Chair of Natural Resources and professor of environmental ethics and philosophy at Oregon State University. He is coeditor (with Kathleen Dean Moore) of *Moral Ground: Ethical Action for a Planet in Peril.*

BRYAN NORTON is Distinguished Professor of Philosophy in the

School of Public Policy at the Georgia Institute of Technology in Atlanta. His newest book is *Sustainable Change.*

STEPHEN J. PYNE is Regents' Professor in the School of Life Sciences at Arizona State University in Tempe. A historian specializing in the study of the environment, fire, and exploration, his most recent book is *Fire: Nature and Culture.*

ANDREW C. REVKIN is an environmental journalist and author. He currently writes the Dot Earth blog for the *New York Times'* Opinion section. He is also Senior Fellow for Environmental Understanding at the Pace Academy for Applied Environmental Studies at Pace University In New York.

HOLMES ROLSTON III is University Distinguished Professor and professor emeritus of philosophy at Colorado State University in Fort Collins. His most recent book is *A New Environmental Ethics: The Next Millennium for Life on Earth.* He has lectured on environmental ethics on all seven continents.

AMY SEIDL is an ecologist, writer, and lecturer in the Environmental Program at the University of Vermont in Burlington. She is author of *Finding Higher Ground: Adaptation in the Age of Warming.*

JACK WARD THOMAS is a wildlife biologist and forester. He was the thirteenth chief of the US Forest Service, serving during the Clinton administration years of 1993–96.

DIANE J. VOSICK is director of policy and partnership in the Ecological Restoration Institute (ERI) at Northern Arizona University in Flagstaff. She works with policy makers, environmental stakeholders, business interests, and land managers to achieve the goal of ecological restoration of federal forest land.

JOHN A. VUCETICH is professor of animal ecology at Michigan Technological University in Houghton. He coleads research on the internationally recognized wolves and moose project of Isle Royale, the longest continuous study of a predator-prey system in the world.

HAZEL WONG is the director of Conservation Campaigns, a division within External Affairs at The Nature Conservancy. She has been on the forefront of researching how to engage communities of color in conservation and advocating for more diverse representation in the conservation community.

DONALD WORSTER is the Hall Distinguished Professor Emeritus at the University of Kansas and Distinguished Foreign Expert at Renmin University of China in Beijing. His most recent book is *A Passion for Nature: The Life of John Muir.*

Index